Protected Cultivation of Horticultural Crops: Advances and Sustainability

Protected Cultivation of Horticultural Crops: Advances and Sustainability

Guest Editors

Weijie Jiang
Wenna Zhang

MDPI

Basel • Beijing • Wuhan • Barcelona • Belgrade • Novi Sad • Cluj • Manchester

Guest Editors
Weijie Jiang
Institute of Vegetables and Flowers
Chinese Academy of Agricultural Sciences
Beijing
China

Wenna Zhang
College of Horticulture
China Agriculture University
Beijing
China

Editorial Office
MDPI AG
Grosspeteranlage 5
4052 Basel, Switzerland

This is a reprint of the Special Issue, published open access by the journal *Plants* (ISSN 2223-7747), freely accessible at: www.mdpi.com/journal/plants/special_issues/2E0L9LR997.

For citation purposes, cite each article independently as indicated on the article page online and using the guide below:

Lastname, A.A.; Lastname, B.B. Article Title. *Journal Name* **Year**, *Volume Number*, Page Range.

ISBN 978-3-7258-3152-4 (Hbk)
ISBN 978-3-7258-3151-7 (PDF)
https://doi.org/10.3390/books978-3-7258-3151-7

Contents

About the Editors vii

Hongjun Yu, Peng Liu, Jingcheng Xu, Tanyu Wang, Tao Lu and Jie Gao et al.
The Effects of Different Durations of Night-Time Supplementary Lighting on the Growth, Yield, Quality and Economic Returns of Tomato
Reprinted from: *Plants* **2024**, *13*, 1516, https://doi.org/10.3390/plants13111516 1

Soheil Fallah, Sasan Aliniaeifard, Mahboobeh Zare Mehrjerdi, Shima Mirzaei and Nazim S. Gruda
Night Interruption with Red and Far-Red Light Optimizes the Phytochemical Composition, Enhances Photosynthetic Efficiency, and Increases Biomass Partitioning in Italian Basil
Reprinted from: *Plants* **2024**, *13*, 3145, https://doi.org/10.3390/plants13223145 15

Hyo Jun Bae, Seong-Hoon Kim, Yuseok Jeong, Sungjin Park, Kingsley Ochar and Youngsin Hong et al.
Optimal Planting Time for Summer Tomatoes (*Lycopersicon esculentum* Mill.) Cropping in Korea: Growth, Yield, and Photosynthetic Efficiency in a Semi-Closed Greenhouse
Reprinted from: *Plants* **2024**, *13*, 2116, https://doi.org/10.3390/plants13152116 33

Jiawei Cui, Yongxue Zhang, Hongmei Zhang, Haijun Jin, Lizhong He and Hong Wang et al.
Low-Potassium Fruits and Vegetables: Research Progress and Prospects
Reprinted from: *Plants* **2024**, *13*, 1893, https://doi.org/10.3390/plants13141893 47

Xianjun Chen, Hongwei Han, Yundan Cong, Xuezhen Li, Wenbo Zhang and Jinxia Cui et al.
Ascorbic Acid Improves Tomato Salt Tolerance by Regulating Ion Homeostasis and Proline Synthesis
Reprinted from: *Plants* **2024**, *13*, 1672, https://doi.org/10.3390/plants13121672 63

Yongdong Qin, Ao Gong, Xigang Liu, Nan Li, Tuo Ji and Jing Li et al.
Testing a Simulation Model for the Response of Tomato Fruit Quality Formation to Temperature and Light in Solar Greenhouses
Reprinted from: *Plants* **2024**, *13*, 1662, https://doi.org/10.3390/plants13121662 81

Huan Liang, Jiangfeng Liu, Xianfeng Shi, Mihong Ge, Juhong Zhu and Dehuan Wang et al.
An Integrated Analysis of Anatomical and Sugar Contents Identifies How Night Temperatures Regulate the Healing Process of Oriental Melon Grafted onto Pumpkin
Reprinted from: *Plants* **2024**, *13*, 1506, https://doi.org/10.3390/plants13111506 97

Theodoros Petrakis, Paraskevi Ioannou, Foteini Kitsiou, Angeliki Kavga, George Grammatikopoulos and Nikos Karamanos
Growth and Physiological Characteristics of Strawberry Plants Cultivated under Greenhouse-Integrated Semi-Transparent Photovoltaics
Reprinted from: *Plants* **2024**, *13*, 768, https://doi.org/10.3390/plants13060768 108

Liangjie Guo, Xinyi Chen, Shiye Yang, Ruimin Zhou, Shenyan Liu and Yanfei Cao
Optimized Design of Irrigation Water-Heating System and Its Effect on Lettuce Cultivation in a Chinese Solar Greenhouse
Reprinted from: *Plants* **2024**, *13*, 718, https://doi.org/10.3390/plants13050718 122

Xu Zhang, Jinxin Peng, Xiaodong Hao, Guifang Feng, Yanhui Shen and Guanghui Wang et al.
Serratia marcescens LYGN1 Reforms the Rhizosphere Microbial Community and Promotes Cucumber and Pepper Growth in Plug Seedling Cultivation
Reprinted from: *Plants* **2024**, *13*, 592, https://doi.org/10.3390/plants13050592 . . . . **137**

About the Editors

Weijie Jiang

He has been engaged in research and extension work on protected horticulture since 1985 and became the PI of the soilless culture research group in IVF/CAAS since 1995. His research areas are related to soilless cultivation for vegetable production under protected cultivation, and he has been appointed as Chairman of the Protected Horticulture Branch, Chinese Society for Horticultural Sciences, since 2013, as well as the Chairman of the Substrates Branch, Chinese Association of Vegetables, since 2018.

He has carried out and successfully completed the FAO/UNDP project entitled Soilless Culture for Vegetable Production and has coordinated several national key projects and MOA (Ministry of Agriculture) projects on protected horticulture. He is the coordinator of the Chinese team of partners in the EU H2020 project (SiEUGreen 774233): Sino-European innovative green and smart cities since 2017; he is also the coordinator of projects in the National Key Research and Development Program of China since 2019. He has been awarded second prizes of the MOA Scien-tech Progress reward in the field of soilless culture in 1996, 2002, 2010, and 2019.

He has published five books and over 100 papers related to protected horticulture and soilless cultivation, either as a first author or corresponding author.

He was invited as a horticultural consultant by the African Development Bank for the Project of Zambian Agricultural Diversification in 2001, by the FAO for the project of green food from green roofs in Egypt in 2002 and 2003, and by CDB for a greenhouse survey in the Caribbean Islands in 2009. He was also invited as co-convener of symposiums related to protected horticulture at the International Horticultural Congress in 2006 (Korea) and 2014 (Australia) and as co-convener of Greensys in 2017 (Beijing).

Wenna Zhang

Dr. Zhang, Wenna, Associate Professor, Supervisor of Ph.D students.

Department of Vegetable Science, College of Horticulture Science, China Agricultural University. Research aspects: high-efficiency cultivation and environmental regulation of protected vegetable crops, factory seedling cultivation, and systemic signaling communication between scion and stock in grafted vegetables.

As the first or corresponding author (including co-authors), Dr. Wenna Zhang has more than 30 SCI articles in academic journals such as *Nature Plants, Plant Cell, New Phytologist, Plant, Cell & Environment, The Plant Journal*, etc. Dr. Wenna Zhang has hosted two National Natural Science Foundation of China projects and has participated in China National Key R&D Plan Projects. Dr. Wenna Zhang is also on the Youth Editorial Board of Vegetable Research and the Editorial Board of BMC Plant Biology.

plants

MDPI

Article

The Effects of Different Durations of Night-Time Supplementary Lighting on the Growth, Yield, Quality and Economic Returns of Tomato

Hongjun Yu [1,†], Peng Liu [1,†], Jingcheng Xu [1,2,†], Tanyu Wang [1], Tao Lu [1], Jie Gao [3], Qiang Li [1,*] and Weijie Jiang [1,*]

1 Institute of Vegetables and Flowers, Chinese Academy of Agricultural Sciences, Beijing 100081, China; yuhongjun@caas.cn (H.Y.); saaslp2023@163.com (P.L.); xujingcheng94@163.com (J.X.); tanyuwang1989@163.com (T.W.); lutao@caas.cn (T.L.)
2 Taizhou Academy of Agricultural Sciences, Taizhou 318014, China
3 College of Horticulture, Xinjiang Agricultural University, Urumqi 830052, China; ofc111@163.com
* Correspondence: liqiang05@caas.cn (Q.L.); jiangweijie@caas.cn (W.J.)
† These authors contributed equally to this work.

Abstract: To achieve higher economic returns, we employ inexpensive valley electricity for night-time supplementary lighting (NSL) of tomato plants, investigating the effects of various durations of NSL on the growth, yield, and quality of tomato. Tomato plants were treated with supplementary light for a period of 0 h, 3 h, 4 h, and 5 h during the autumn–winter season. The findings revealed superior growth and yield of tomato plants exposed to 3 h, 4 h, and 5 h of NSL compared to their untreated counterparts. Notably, providing lighting for 3 h demonstrated greater yields per plant and per trough than 5 h exposure. To investigate if a reduced duration of NSL would display similar effects on the growth and yield of tomato plants, tomato plants received supplementary light for 0 h, 1 h, 2 h, and 3 h at night during the early spring season. Compared to the control group, the stem diameter, chlorophyll content, photosynthesis rate, and yield of tomatoes significantly increased upon supplementation with lighting. Furthermore, the input–output ratios of 1 h, 2 h, and 3 h NSL were calculated as 1:10.11, 1:4.38, and 1:3.92, respectively. Nonetheless, there was no detectable difference in yield between the 1 h, 2 h, and 3 h NSL groups. These findings imply that supplemental LED lighting at night affects tomato growth in the form of light signals. Night-time supplemental lighting duration of 1 h is beneficial to plant growth and yield, and its input–output ratio is the lowest, which is an appropriate NSL mode for tomato cultivation.

Keywords: tomato; night-time supplemental light; yield; signal; economic returns

Citation: Yu, H.; Liu, P.; Xu, J.; Wang, T.; Lu, T.; Gao, J.; Li, Q.; Jiang, W. The Effects of Different Durations of Night-Time Supplementary Lighting on the Growth, Yield, Quality and Economic Returns of Tomato. *Plants* **2024**, *13*, 1516. https://doi.org/10.3390/plants13111516

Academic Editor: Giorgio Perrella

Received: 8 March 2024
Revised: 2 May 2024
Accepted: 27 May 2024
Published: 31 May 2024

1. Introduction

In light of a recent investigation, solar greenhouses have become the most important facilities in protected vegetable production in northern China. Light deficiency is one of the critical environmental limiting factors affecting crop yield for the greenhouse vegetable production [1]. Light is a key environmental variable, whose effects on plant growth and development are mainly reflected in three aspects, including light intensity [2], light quality [3,4] and illumination time [5], which can regulate plant morphogenesis, photosynthesis [6], material metabolism and yield [7,8]. Poor irradiance is one of the major limiting factors for plant growth and development, especially for vegetative growth. Some papers have described that plants under low-light conditions suffer from improper metabolism and senescence acceleration [9,10]. Low solar radiation is harmful to the photosynthesis of rice, which hinders the accumulation and distribution of dry matter, resulting in a lower yield and poor quality of rice [11]. Artificial regulation and optimization of light conditions can improve the photosynthetic characteristics [12], promote crop growth [13], increase the yield [14], enhance the quality, and effectively delay the organ senescence of plants [15]. Some researchers have verified that light supplementation by LED can enhance the photosynthetic capacity of cucumber [16], and at bloom time, supplementary illumination can markedly promote the growth of cucumber fruits [17].

In recent years, LED supplementary lighting technology has been commonly used in protected agricultural production. Numerous studies have confirmed that LED lighting supplementation can be beneficial to crop growth and development, increase yield and improve quality [15,18]. Previously, several meaningful studies about the effects of light intensity and light quality on the performance of different plants were conducted [19,20], but there is little research available to elucidate the effects of NSL duration on tomato growth, especially studies on the effects of NSL duration on the whole growth cycle of tomato and the corresponding economic returns.

To enhance the energy resource utilization efficiency and incentivize electricity consumers to allocate electrical power consumption rationally, utility firms segment 24 h per day into peak (8:00–22:00) and off-peak (22:00–8:00) intervals, offering varying electricity rates during these respective timescales. During our experimental setup, we circumvent the most heavily utilized interval of electric power consumption (i.e., the 'peak' period), instead opting to initiate illumination once the cost of electricity plummets (i.e., the 'valley' period). The primary aim of this study is to investigate the influence of night-time supplementary light (NSL) on tomato development and fruitfulness throughout various developmental phases, as well as assess its impact on tomato yield and potential monetary rewards. No conclusive association was observed between the length of additional lighting and tomato yield, despite previous findings from daylight studies suggesting that such treatment's advantages escalated alongside extended exposure durations [21,22]. Consequently, we aimed to verify if the enhancement in crop growth and output induced by nocturnal supplementary illumination correlates with lighting duration. As a means to delve deeper into this phenomenon, we reduced the elongating photoperiod, adding an experiment encompassing supplementary lighting durations of 1 h, 2 h, and 3 h. Our investigations served to thoroughly evaluate the effects of NSL upon tomato development, identify the optimal lighting duration, and establish a solid foundation for developing a low-cost yet highly advantageous nocturnal supplementary lighting technique capable of widespread implementation.

2. Materials and Methods

2.1. Experimental Materials

The tomato variety DER0899 was selected as the plant material in this experiment. Tomato seeds were successfully germinated and sown in 32-plug trays on 18 July 2016 and 10 January 2017, respectively. Seedlings reaching the four true-leaf stage (10 August 2016 and 22 February 2017, respectively) were transplanted into cultivation troughs (7 m long × 0.35 m wide × 0.15 m deep) and each trough had 39 tomato seedlings. The cultivation substrate was cocopeat, provided by Beijing Yinong Agricultural Technology Company, Beijing, China. The organic fertilizer used in this trial was produced by Qian'an Fusheng Animal Husbandry Technology Company, Tangshan, Hebei, China. This organic fertilizer was blended with coconut coir at a rate of 55 kg:1 m^3 (w/v). The drip irrigation system used an integrated precision automatic control system provided by Beijing Zigvine Technology Company, Beijing, China. With soil humidity ≤ 40% and light intensity ≥ 450 $\mu mol \cdot m^{-2} \cdot s^{-1}$, the drip irrigation system started and ran for four minutes. The planting density was 4 plants per square meter. The formula of the nutrient solution used in the experiment is the special Yamazaki tomato formula (Table 1) [23].

Table 1. Composition of special nutrient solution for tomato proposed by Yamazaki.

Macronutrients	Final Concentration (mg/L)	Micronutrients	Final Concentration (mg/L)
$Ca(NO_3)_2{\cdot}4H_2O$	614	EDTA-FeNa$\cdot 3H_2O$	6.4
KNO_3	430	$MnSO_4{\cdot}H_2O$	1.7
$NH_4H_2PO_4$	267	$ZnSO_4{\cdot}7H_2O$	1.5
$(NH_4)_2SO_4$	33	$Na_2B_4O_7{\cdot}8H_2O$	4.8
$MgSO_4{\cdot}7H_2O$	430	$CuSO_4{\cdot}5H_2O$	0.2
K_2SO_4	397	$Na_2MoO_4{\cdot}2H_2O$	0.2

LED tubes (97 cm × 10 cm) used as supplemental lighting sources were obtained from Chongqing Xinglian Yunke Technology Development Co., Ltd, Chongqing, China. The total power of each LED tube was 70W, and the red–blue light ratio was 2:1, producing an estimated photosynthetically active photon flux density (PPFD) of approximately 88 $\mu mol{\cdot}m^{-2}{\cdot}s^{-1}$ when measured at a distance of 10cm from the LED module. For each cultivation trough, four LED tubes were installed vertically above the plant canopy at a height of around 50 cm, with the distances between the LED tubes and the plant canopy adjusted based on the plant's growth and kept constant during the entire experiment. The measured light intensity reaching the plant canopy was approximately 30 $\mu mol{\cdot}m^{-2}{\cdot}s^{-1}$.

2.2. Experimental Design

The NSL trial of the autumn–winter season (AW) tomato was conducted in the solar greenhouse of the Xiaotangshan special vegetable base. The experiment consisted of 4 treatments, including control (without lighting supplementation), supplemental lighting for 0, 3 h (23:00~02:00), 4 h (23:00~03:00), 5 h (23:00~04:00). To tomato plants with six true leaves, supplemental lighting at night was applied from 17 August 2016 to 30 October 2016. The experiment consisted of three replicates and the total number of the four treatments was 468, and each cultivation trough was used as a replication. The average sunlight exposure time in the greenhouse was 8 h per day, with an average light intensity of 250 $\mu mol{\cdot}m^{-2}{\cdot}s^{-1}$. The temperature inside the greenhouse was 23–26 °C and 10–12 °C during the day and night, respectively.

In order to further explore the effect of the duration of NSL on tomato production, the NSL trial of the early spring (ES) season tomato was conducted in the solar greenhouse of the Xiaotangshan special vegetable base. The experiment consisted of 4 treatments, including control (without lighting supplementation), supplemental lighting for 1 h (23:00~00:00), 2 h (23:00~01:00), 3 h (23:00~02:00). To tomato plants with six true leaves, supplemental lighting at night was applied from 28 February 2017 to 14 May 2017. The experiment consisted of three replicates and the total number of the four treatments was 468, and each trough was used as a replication. The average sunlight exposure time in the greenhouse was 10 h per day, with an average light intensity of 300 $\mu mol{\cdot}m^{-2}{\cdot}s^{-1}$. The temperature inside the greenhouse was 26–32 °C and 15–18 °C during the day and night, respectively.

Black clothes were used for blocking the light from adjacent troughs of other treatments at night, and they were removed during the day. The drip irrigation system, a water- and fertilizer-integrated precision automatic control system that could automatically irrigate based on substrate humidity and light intensity, was employed in this trial. After completion of these trials, the values of each electric meter were recorded for use in calculating the energy consumption per tank.

2.3. *Measurements and Methods*

2.3.1. Plant Growth

On days 15, 30, 45, 60, and 75 following the illumination treatment (DAT), plant height, stem diameter (measured using a scale with the fifth leaf beneath the apex closed), and chlorophyll content were measured from 18 randomly selected plants from each group during each observation period using ruler, electronic vernier caliper, and SPAD-502 Plus chlorophyll meter, respectively.

2.3.2. Leaf Net Photosynthesis Rate

At 15, 45, and 75 days post initiation of supplementary lighting, the net photosynthesis rate for three independently growing tomato plants per experimental replicate was assessed using a portable LI-6400 photosynthesis system (LI-6400XT, Li-COR, Inc., Lincoln, NE, USA) as described previously [24]. The parameters used were as follows: photosynthetic photon flux density, 1000 $\mu mol \cdot m^{-2} \cdot s^{-1}$; and air flow rate inside the sample chamber, 400 $mol \cdot s^{-1}$.

2.3.3. Fruit Yield

The yield per trough for each treatment was calculated by the cumulative yield throughout the harvest time, yield per plant, and single-fruit weight of 18 selected tomato plants weighed by an electronic balance.

2.3.4. Fruit Quality

Ripe tomato fruits with uniform maturity were randomly selected from each treatment to measure the nutritional quality parameters, including titratable acid content, soluble sugar content, Vitamin C content and soluble solids' content, determined by titration with sodium hydroxide, anthrone-sulfuric acid colorimetry, spectrophotometer colorimetry and portable Abbe refractometer, respectively.

2.4. *Statistical Analysis*

All data were analyzed using Excel 2003 and SPSS 17.0 software, and the statistical significance of the differences between treatments was determined by Duncan's multiple range test ($p < 0.05$). Different letters indicate significant differences.

3. Results

3.1. *Effects of NSL Duration on Plant Height of Tomato*

In the autumn–winter season, the plant height of tomatoes cultivated under 3 h NSL was strikingly higher than that of other treatments. And there was no significant difference between the plant height of tomatoes cultivated under 4 h and that of the control. However compared with the control, the plant height of tomatoes under the 5 h supplementary lighting treatment decreased markedly 30 and 60 days after the lighting treatments (DAT) and no difference was observed at other times (Figure 1A).

In the early spring season, the tomato plants to which artificial night-time lighting for 1 h was applied were shorter than the plants without lighting supplementation after 15 days. However, there was no difference among the four treatments 30 DAT. Additionally the tomatoes receiving the 2 h supplemental lighting treatment were obviously taller than those of other groups on 45 DAT, but the plant height of lighting treatments increased from 60 to 75 DAT, and the plant height of the 2 h night-time lighting supplementation group was markedly lower than that of the other treatments. Moreover, there was no difference in tomato plant height between those subjected to lighting for under 3 h and those without lighting (Figure 1B).

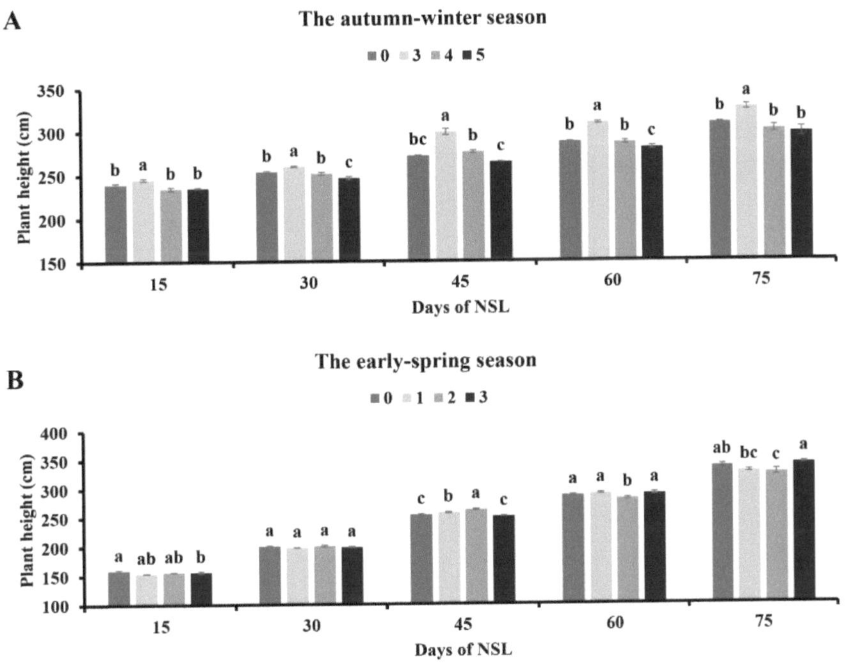

Figure 1. Effects of NSL duration on plant height of tomato in the autumn–winter season (**A**) and the early spring season (**B**). Different letters indicate statistically significant differences ($p < 0.05$).

3.2. Effects of NSL Duration on Stem Diameter of Tomato

In the autumn–winter season, compared to the control, the exposure of tomato plants to supplementary lighting contributed to a marked increase in stem diameter from 15 to 60 DAT. The tomato stem diameter of the 3 h lighting treatment was remarkably increased, except on 75 DAT. And the tomato stem diameter of the 4 h and 5 h lighting treatments was increased significantly except on 30 and 75 DAT. Among different lighting treatments, the stem diameter of tomato was similar 15 DAT, but the group in which supplemental lighting for 3 h was used had a thicker stem diameter than that with 5 h lighting 30 DAT. In addition, the group of supplemental lighting for 3 h showed better behavior than the 4 h and 5 h groups regarding the tomato stem diameter from 45 to 75 DAT (Figure 2A).

In the early spring season, compared to the control, the three supplementary illumination treatments could remarkably increase the stem diameter, except on 60 DAT. The tomato stem diameter of the 3 h of NSL group increased strikingly 60 DAT compared to that of the control. Among the various durations of the NSL treatments, exposing tomato plants to 1 h NSL showed a better result than 2 h and 3 h on stem diameter 15 DAT. Tomato plants treated with supplemental lighting for 1 h and 2 h had a thicker stem diameter than 3 h 30 DAT. No significant difference in tomato stem diameter was noted among those lighting supplementation treatments on 45 and 60 DAT. The stem index was remarkably higher by the addition of lighting for 3 h than 1 h and 2 h, but a non-significant difference was detected between the treatments with lighting for 1 h and 2 h at 75 DAT (Figure 2B).

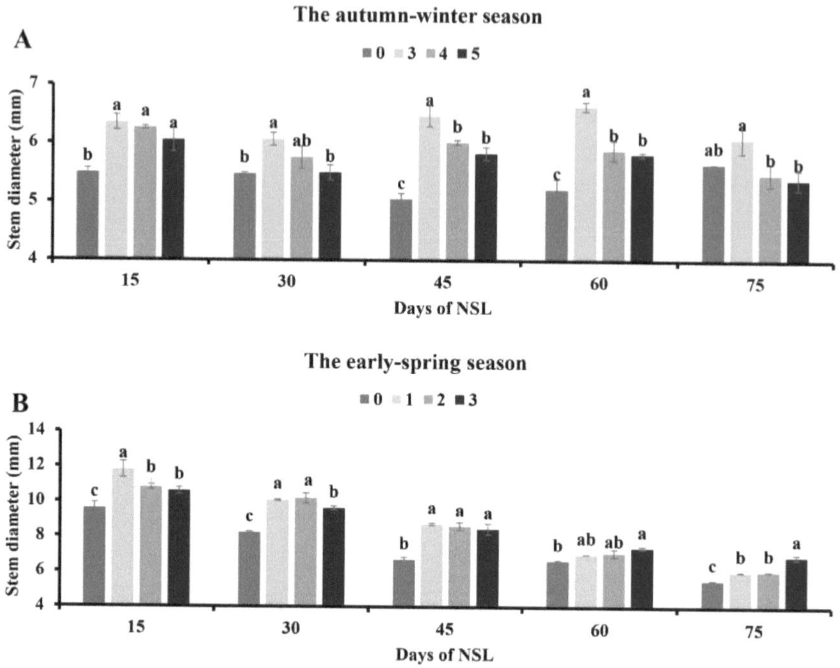

Figure 2. Effects of NSL duration on stem diameter of tomato plants in the autumn–winter season (**A**) and the early spring season (**B**). Different letters indicate statistically significant differences ($p < 0.05$).

3.3. *Effects of NSL Duration on Chlorophyll Content*

In the autumn–winter season, compared to the control, all three night-time supplemental lighting treatments could obviously enhance the chlorophyll content presented in tomato leaves on 15 and 45 DAT, but there was no significant difference among the three lighting groups. A total of 30 DAT, the chlorophyll content of tomato plants receiving 3 h and 5 h night-time lighting supplementation tended to be markedly higher than that of plants without supplementing lighting. And the 5 h supplementary illumination treatment led to significantly higher chlorophyll content than the 3 h lighting treatment, while no difference was observed between the 3 h and 4 h lighting treatments. The chlorophyll content was similar in the absence or presence of supplemental illumination 60 DAT (Figure 3A).

In the early spring season, the chlorophyll content of the supplemental lighting groups was significantly increased compared to that of the control from 15 to 75 DAT. Groups utilizing night-time lighting supplementation had the same effect on the SPAD value. The chlorophyll content of tomato plants cultured under 1 h night-time lighting tended to be markedly higher than that of the other two lighting treatments, and there was no remarkable difference between the other two lighting groups on 30 DAT. A total of 60 DAT, tomato plants exposed to 1 h lighting supplementation contained a markedly higher chlorophyll content relative to the other two supplemental lighting treatments, and the 3 h lighting group had a lower chlorophyll content than the 2 h lighting group (Figure 3B).

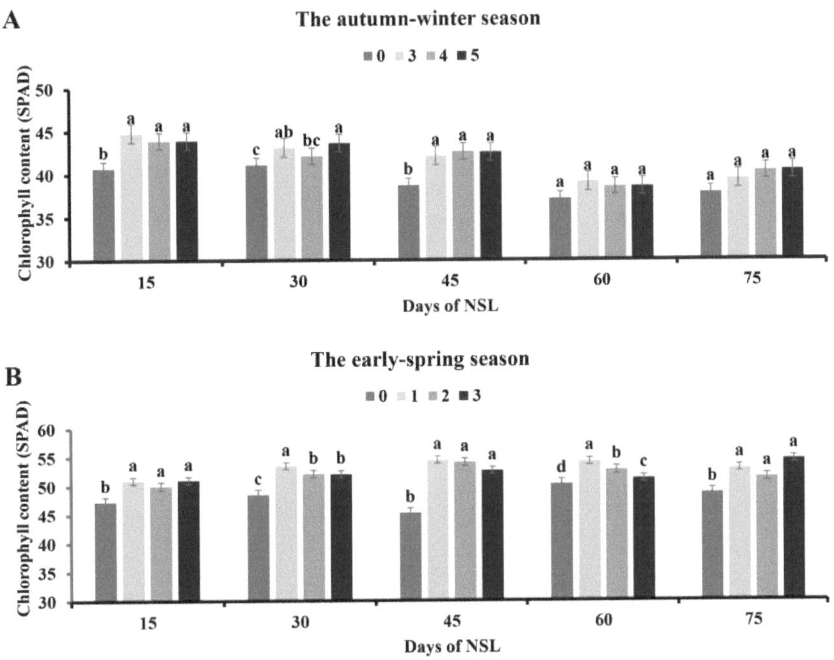

Figure 3. Effects of NSL duration on the chlorophyll content of tomato leaves in the autumn–winter season (**A**) and the early spring season (**B**). Different letters indicate statistically significant differences ($p < 0.05$).

3.4. Effects of NSL Duration on Net Photosynthetic Rate

In the autumn–winter season, the net photosynthetic rate of the supplemental lighting groups was significantly enhanced compared to that of the control while no obvious difference was noted among the lighting supplementation treatments on 15 DAT. The net photosynthetic rate of the tomato plants with 5 h lighting was markedly higher than that of the other groups 45 DAT, while the other three groups had a consistent effect on the photosynthetic rate. On 75 DAT, the net photosynthetic rate of tomato plants grown under 4 h night-time lighting supplementation tended to be obviously higher than the other treatments', but there was no remarkably different treatment effect between the control and the other two lighting groups (Figure 4A).

In the early spring season, it can be observed that compared with the control, the net photosynthetic rate enhanced obviously in all different lighting supplementation treatment groups. No significant difference in the net photosynthetic rate was noted between the three lighting treatments on 15 and 45 DAT. Supplementing lighting for 2 h and 3 h had the same influence of photosynthetic capability, and the 1 h lighting treatment had a better effect than the other two lighting treatments on the photosynthetic parameter (Figure 4B).

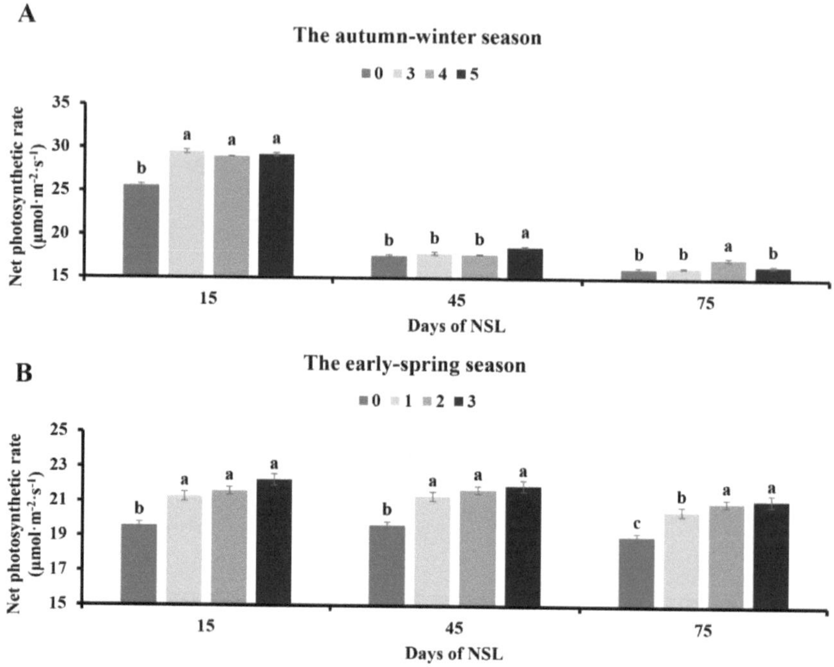

Figure 4. Effects of NSL duration on net photosynthetic rate of the autumn–winter season tomato (**A**) and the early spring season tomato (**B**). Different letters indicate statistically significant differences ($p < 0.05$).

3.5. Effects of NSL Duration on Tomato Yield and Input–Output Ratio

In the autumn–winter season, compared with the no additional lighting group, the single-fruit weight, yield per plant and yield per trough of the lighting supplementation groups increased significantly; however, those night-time supplemental lighting treatments exerted an equivalent influence on single-fruit weight. Supplementing lighting for 3 h showed better performances than 5 h on yield per plant and yield per trough. Among the three lighting groups, the 3 h supplemental lighting group had the least electricity input accompanied by the highest increased yield per trough and the biggest input–output ratio 1:6.37 (Table 2).

In the early spring season, the treatment of 3 h NSL led to significantly higher single-fruit weight than that of the control, while the other two supplemental lighting treatments matched up with the control on this index. Among the three lighting groups, no marked difference was found in single-fruit weight. The yield per plant and the yield per trough of all lighting addition treatments were significantly higher than those of the control. Although the electricity consumption was different among the three lighting supplementation groups, there was no significant difference in yield per plant and yield per trough. In summary, supplementing lighting for 1 h had the biggest input–output ratio, 1:10.11 (Table 2).

The input of electricity charge was CNY 0.3 per kWh, and the output of tomato fruit was CNY 4 per kilogram. The one-off input of LED tubes was not included in this table. Different letters indicate statistically significant differences ($p < 0.05$).

Table 2. Effects of NSL duration on tomato yield and input–output ratio.

NSL Duration (Hours)	Single-Fruit Weight (g)	Yield per Plant (kg)	Yield per Trough (kg)	Increased Yield per Trough (kg)	Power Consumption per Trough (kWh)	Power Input per Trough (CNY)	Increased Output (CNY)	Input-Output Ratio
	The Autumn–Winter Season							
0	120 ± 2.0 b	2.68 ± 0.06 c	104.60 ± 2.50 c	-	-	-	-	-
3	133 ± 0.6 a	3.46 ± 0.09 a	134.94 ± 3.42 a	30.34	63.55	19.07	121.36	1:6.37
4	130 ± 1.0 a	3.21 ± 0.05 ab	125.24 ± 1.91 ab	20.65	85.50	25.65	82.60	1:3.22
5	128 ± 0.9 a	3.13 ± 0.13 b	122.04 ± 5.14 b	17.44	106.35	31.91	69.76	1:2.19
	The early spring season							
0	106 ± 4.0 b	2.45 ± 0.23 b	95.49 ± 5.27 b	-	-	-	-	-
1	114 ± 1.0 ab	2.91 ± 0.04 a	113.68 ± 0.84 a	18.20	24.00	7.20	72.80	1:10.11
2	108 ± 2.0 ab	2.86 ± 0.07 a	111.41 ± 1.52 a	15.93	48.50	14.55	63.72	1:4.38
3	117 ± 2.0 a	2.98 ± 0.03 a	116.05 ± 0.69 a	20.56	70.00	21.00	82.24	1:3.92

3.6. Effects of NSL Duration on Fruit Quality of Tomato

In the autumn–winter season, the concentration of Vitamin C and soluble solids in tomato fruits under night-time supplemental LED lighting for 3 h was significantly greater than that of the control, while titratable acidity was markedly lower than that of the control. The soluble sugar content of tomato fruits irradiated with 3 h supplementary illumination seemed to be higher than that of plants subjected to darkness at night but the difference was not significant between the two groups. Compared with the control, the content of titratable acid and Vitamin C in tomato fruits exposed to 4 h night-time lighting supplementation was markedly higher, and the content of soluble sugar and soluble solids was also higher than that of the control but the difference did not reach a significant level. The tomato fruits of the 5 h night-time supplementary illumination group stored remarkably more titratable acid, soluble sugar and soluble solids than those of the control, while no remarkable difference in Vitamin C content was found between them. The 3 h supplemental lighting application group had the best performance on the acid–sugar ratio and total soluble solids–acid ratio among the four treatments (Table 3).

Table 3. Effects of NSL duration on fruit quality of tomato.

NSL Duration (Hours)	The Autumn–Winter Season					
	Titratable Acidity (%)	**Soluble Sugar (%)**	**Vitamin C (mg/100 g)**	**Total Soluble Solids (%)**	**Acid-Sugar Ratio**	**Total Soluble Solids-Acid Ratio**
0	0.25 ± 0.002 b	1.57 ± 0.08 b	16.14 ± 0.84 c	2.93 ± 0.07 c	6.24	11.67
3	0.22 ± 0.002 c	1.64 ± 0.07 ab	19.54 ± 0.34 ab	3.37 ± 0.07 b	7.33	15.00
4	0.27 ± 0.00 a	1.58 ± 0.09 b	21.24 ± 0.87 a	3.03 ± 0.03 c	5.75	11.04
5	0.28 ± 0.006 a	1.87 ± 0.02 a	17.22 ± 0.77 bc	3.97 ± 0.03 a	6.55	13.93
	The early spring season					
0	0.62 ± 0.007 a	1.92 ± 0.04 b	21.54 ± 0.50 b	5.00 ± 0.00 c	3.10	8.12
1	0.54 ± 0.02 b	2.03 ± 0.08 b	21.90 ± 0.49 b	5.17 ± 0.03 b	3.77	9.59
2	0.63 ± 0.006 a	2.49 ± 0.04 a	26.27 ± 0.08 a	6.17 ± 0.03 a	3.94	9.74
3	0.49 ± 0.005 b	1.95 ± 0.009 b	21.56 ± 0.20 b	5.03 ± 0.03 c	3.98	10.27

Different letters indicate statistically significant differences ($p < 0.05$).

In the early spring season, the 1 h lighting treatment improved the content of soluble solids compared to the control and the 3 h lighting group; in addition, titratable acidity was significantly lower than that of the control, and the content of Vitamin C and soluble sugar was found to be insignificant compared with that of the control. The content of soluble sugar, Vitamin C and soluble solids accumulated in tomato fruits exposed to 2 h supplementary illumination was markedly modified compared tot to that of the other treatments. The content of titratable acid of tomato fruits exposed to 2 h night-time lighting supplementation was significantly higher than that of the 1 h and 3 h lighting treatments but was insignificant compared with that of the control. The tomato fruits of the 3 h night-time supplementary illumination group contained considerably less titratable acid than those of the control but had an equal level of soluble sugar, Vitamin C and soluble solids compared with the control. The control had a lower acid–sugar ratio and total soluble solids–acid ratio than all supplementary illumination treatments, indicating that supplemental lighting at night had a positive effect on the quality improvement of tomato fruits (Table 3).

4. Discussion

4.1. NSL Affects Tomato Production in the Form of Light Signals

With the wide application of LED lights, it is possible to supplement lighting in plant factories and facilities [25]. Recently, numerous studies have examined the effects of various factors related to supplementary lighting duration, quality [26], and intensity [27] on plant growth during the day. However, there is only a limited number of studies examining the impact of nocturnal supplementary lighting on plant development. In our study, we employed night-time supplementary lighting technology to disrupt the dark cycle of plants. Additionally, the intensity of supplementary lighting used in this experiment at night was

30 $\mu mol\cdot m^{-2}\cdot s^{-1}$, which did not supply sufficient energy for photosynthesis in tomatoes. Notably, our findings indicate that nocturnal supplementary lighting can substantially enhance the photosynthetic rate in the daytime, yield, and quality of tomatoes. Thus, we hypothesize that nocturnal supplementary lighting may serve as a type of light signal promoting plant growth.

Nocturnal supplementary lighting disrupts the circadian rhythms of plants, which can influence multiple aspects of plant physiology [28,29]. Additionally, Velez-Ramirez et al. found that the expression of CAB-13 (III light harvesting chlorophyll a/b binding protein 13) is regulated by the circadian rhythm, and this protein can regulate the effects of photosystem I and II light harvesting, which is an important factor affecting plant photosynthesis [30]. The results of our experiment show that supplemental lighting at night can improve the photosynthetic capacity during the whole production period of tomato in the early spring season and the early stage of tomato in the autumn–winter season (Figure 4), and that is consistent with the report by Tewolde et al. [31]. The chlorophyll content and distribution in plant leaf tissue also affects photosynthetic capacity. It has been reported that supplemental lighting at night increases the chlorophyll content in the middle and lower canopy of tomato [31]. This experiment also found that supplemental lighting at night increased the chlorophyll content of tomato (Figure 3). The increase in chlorophyll content may be one of the reasons why supplemental lighting at night promotes photosynthesis during the day.

Assimilate partitioning plays a crucial role in plant growth and morphological development [32]. Light qualities and photoperiods affect how assimilates distribute within plants [33,34]. Whether night-time supplementary lighting, acting as light signals, regulates the distribution of assimilates remains to be further investigated. These findings offer novel insights and approaches for regulating environmental light in plant factories and facilities.

4.2. The Duration of Night-Time Supplemental LED Has No Effect on Tomato Yield

In the present research, the findings of the night-time lighting supplementation experiment in autumn suggest that supplementing lighting for 3 h, 4 h and 5 h at night could promote the growth of tomato plants at the earlier growth stage and significantly increase the yield of tomato. However, supplementing lighting for 3 h showed better performances than 5 h on the yield per plant and yield per trough. To further refine the duration and verify our hypothesis that improved plant growth and yield through night-time supplementation might not necessarily correlate with the length of NSL, we reduced the durations from 3 h, 4 h, and 5 h to 1 h, 2 h, and 3 h during the early spring test, respectively. Nonetheless, there was no significant difference in single-fruit weight, yield per plant or yield per trough among the three lighting groups in the early spring season.

Interestingly, we discovered that extended exposure to night-time supplemental lighting does not guarantee an increased tomato yield. Under these conditions, the maximum yield and the highest input–output ratio were achieved when tomato plants received supplementary lighting for 1 h at night. This is inconsistent with many previous reports in the literature [35]. The rationale behind this phenomenon lies in the fact that the night-time supplementary lighting employed in this study functions as light signals to disrupt the tomato's dark cycle. Moreover, excessive supplementary lighting can lead to both energy wastage and detrimental effects on plants [36,37]. One possible explanation for this finding is the carbon imbalance induced by persistent supplementary lighting, which hindered photosynthesis, leaf senescence, and photodamage in tomatoes [38]. Moreover, numerous photoreceptors are involved not only in sensing and responding to light, but also in temperature cues [39]. In previous studies, short days and high temperature accelerated the growth of plants [40]. We found that the effect of the NSL trial in early spring is better than that in the autumn–winter season, which is likely due to the higher temperature in the early spring greenhouse than in the autumn and winter greenhouse. Subsequently, we plan to shorten the duration of night-time supplemental lighting even further to achieve a more effective tomato night-time supplementary lighting pattern.

Using LED to improve the lighting conditions during the process of tomato growth and development is a feasible and efficient method to improve the quality of tomato fruit. Li et al. found that supplementing light in the morning promoted the accumulation of vitamin C, organic acids and sugar [5]. However, we found that the content of soluble sugar, Vitamin C and soluble solids accumulated in tomato fruits subjected to 2 h supplementary illumination was markedly modified compared to that of other treatments in early spring and the content of titratable acid and Vitamin C in tomato fruits exposed to 4 h night-time lighting supplementation was markedly higher, but the content of soluble sugar in tomato fruits exposed to 5 h night-time lighting supplementation was markedly higher in autumn–winter season. Further research is needed to explain this phenomenon.

5. Conclusions

In this experiment, tomato plants were irradiated at night with red and blue dichromatic light at an intensity of 30 μmol·m^{-2}·s^{-1}, and the results indicated that night-time lighting supplementation can promote plant growth. Therefore, we speculate that night-time supplementary lighting affects tomato production in the form of light signals. Interestingly, the duration of night-time supplemental LED has no effect on tomato yield. The total yield per plant, the yield per trough and the input–output ratio were the highest in the treatments with 1 h of supplemental lighting at night in spring stubble. In order to obtain the maximum economic returns, supplementing lighting for 1 h at night should be selected for tomato cultivation, which has the highest input–output ratio in all supplemental lighting treatments.

Author Contributions: H.Y., P.L. and J.X., conceptualization, data curation, methodology, investigation, roles, and writing—original draft. H.Y., P.L., T.W., T.L. and J.G., data curation, formal analysis, investigation, and methodology. H.Y., P.L., J.X., Q.L. and W.J., conceptualization, writing—review and editing, supervision, formal analysis, and funding support. All authors have read and agreed to the published version of the manuscript.

Funding: This research was funded by the National Key Research and Development Program of China (grant No. 2023YFD2300700) and the earmarked fund for the China Agricultural Research System (CARS-23-B-07).

Data Availability Statement: Data are contained within the article.

Conflicts of Interest: The authors declare that the research was conducted in the absence of any commercial or financial relationships that could be construed as potential conflicts of interest.

References

1. Li, Q.; Chai, L.; Tong, N.; Yu, H.; Jiang, W. Potential Carbohydrate Regulation Mechanism Underlying Starvation-Induced Abscission of Tomato Flower. *Int. J. Mol. Sci.* **2022**, *23*, 1952. [CrossRef] [PubMed]
2. Szechyńska-Hebda, M.; Karpiński, S. Light intensity-dependent retrograde signalling in higher plants. *J. Plant Physiol.* **2013**, *170*, 1501–1516. [CrossRef] [PubMed]
3. Douma, J.C.; de Vries, J.; Poelman, E.H.; Dicke, M.; Anten, N.P.; Evers, J.B. Ecological significance of light quality in optimizing plant defence. *Plant Cell Environ.* **2019**, *42*, 1065–1077. [CrossRef] [PubMed]
4. He, R.; Wei, J.; Zhang, J.; Tan, X.; Li, Y.; Gao, M.; Liu, H. Supplemental Blue Light Frequencies Improve Ripening and Nutritional Qualities of Tomato Fruits. *Front. Plant Sci.* **2022**, *13*, 888976. [CrossRef] [PubMed]
5. Wang, S.; Jin, N.; Jin, L.; Xiao, X.; Hu, L.; Liu, Z.; Wu, Y.; Xie, Y.; Zhu, W.; Lyu, J.; et al. Response of Tomato Fruit Quality Depends on Period of LED Supplementary Light. *Front. Nutr.* **2022**, *9*, 833723. [CrossRef] [PubMed]
6. Ren, M.; Liu, S.; Tang, C.; Mao, G.; Gai, P.; Guo, X.; Zheng, H.; Tang, Q. Photomorphogenesis and Photosynthetic Traits Changes in Rice Seedlings Responding to Red and Blue Light. *Int. J. Mol. Sci.* **2023**, *24*, 11333. [CrossRef] [PubMed]
7. de Wit, M.; Galvão, V.C.; Fankhauser, C. Light-Mediated Hormonal Regulation of Plant Growth and Development. *Annu. Rev. Plant Biol.* **2016**, *67*, 513–537. [CrossRef] [PubMed]
8. Paponov, M.; Kechasov, D.; Lacek, J.; Verheul, M.J.; Paponov, I.A. Supplemental Light-Emitting Diode Inter-Lighting Increases Tomato Fruit Growth Through Enhanced Photosynthetic Light Use Efficiency and Modulated Root Activity. *Front. Plant Sci.* **2019**, *10*, 1656. [CrossRef] [PubMed]

9. Kucharewicz, W.; Distelfeld, A.; Bilger, W.; Müller, M.; Munné-Bosch, S.; Hensel, G.; Krupinska, K. Acceleration of leaf senescence is slowed down in transgenic barley plants deficient in the DNA/RNA-binding protein WHIRLY1. *J. Exp. Bot.* **2017**, *68*, 983–996. [CrossRef]
10. Zhu, H.; Li, X.; Zhai, W.; Liu, Y.; Gao, Q.; Liu, J.; Ren, L.; Chen, H.; Zhu, Y. Effects of low light on photosynthetic properties, antioxidant enzyme activity, and anthocyanin accumulation in purple pak-choi (*Brassica campestris* ssp. Chinensis Makino). *PLoS ONE* **2017**, *12*, e179305. [CrossRef]
11. Gad, A.G.; Habiba; Zheng, X.; Miao, Y. Low Light/Darkness as Stressors of Multifactor-Induced Senescence in Rice Plants. *Int. J. Mol. Sci.* **2021**, *22*, 3936. [CrossRef] [PubMed]
12. Monostori, I.; Heilmann, M.; Kocsy, G.; Rakszegi, M.; Ahres, M.; Altenbach, S.B.; Szalai, G.; Pál, M.; Toldi, D.; Simon-Sarkadi, L.; et al. LED Lighting—Modification of Growth, Metabolism, Yield and Flour Composition in Wheat by Spectral Quality and Intensity. *Front. Plant Sci.* **2018**, *9*, 605. [CrossRef] [PubMed]
13. Zou, T.; Huang, C.; Wu, P.; Ge, L.; Xu, Y. Optimization of Artificial Light for Spinach Growth in Plant Factory Based on Orthogonal Test. *Plants* **2020**, *9*, 490. [CrossRef]
14. Samuolienė, G.; Sirtautas, R.; Brazaitytė, A.; Duchovskis, P. LED lighting and seasonality effects antioxidant properties of baby leaf lettuce. *Food Chem.* **2012**, *134*, 1494–1499. [CrossRef]
15. Wang, S.; Fang, H.; Xie, J.; Wu, Y.; Tang, Z.; Liu, Z.; Lv, J.; Yu, J. Physiological Responses of Cucumber Seedlings to Different Supplemental Light Duration of Red and Blue LED. *Front. Plant Sci.* **2021**, *12*, 709313. [CrossRef] [PubMed]
16. Hamedalla, A.M.; Ali, M.M.; Ali, W.M.; Ahmed, M.A.A.; Kaseb, M.O.; Kalaji, H.M.; Gajc-Wolska, J.; Yousef, A.F. Increasing the performance of cucumber (*Cucumis sativus* L.) seedlings by LED illumination. *Sci. Rep.* **2022**, *12*, 852. [CrossRef] [PubMed]
17. Trouwborst, G.; Oosterkamp, J.; Hogewoning, S.W.; Harbinson, J.; van Ieperen, W. The responses of light interception, photosynthesis and fruit yield of cucumber to LED-lighting within the canopy. *Physiol. Plant.* **2010**, *138*, 289–300. [CrossRef] [PubMed]
18. Ntagkas, N.; de Vos, R.C.H.; Woltering, E.J.; Nicole, C.C.S.; Labrie, C.; Marcelis, L.F.M. Modulation of the Tomato Fruit Metabolome by LED Light. *Metabolites* **2020**, *10*, 266. [CrossRef]
19. Zhang, Y.; Hu, W.; Peng, X.; Sun, B.; Wang, X.; Tang, H. Characterization of anthocyanin and proanthocyanidin biosynthesis in two strawberry genotypes during fruit development in response to different light qualities. *J. Photochem. Photobiol. B Biol.* **2018**, *186*, 225–231. [CrossRef]
20. Hanssens, J.; DE Swaef, T.; Steppe, K. High light decreases xylem contribution to fruit growth in tomato. *Plant Cell Environ.* **2015**, *38*, 487–498. [CrossRef]
21. Dannehl, D.; Schwend, T.; Veit, D.; Schmidt, U. Increase of Yield, Lycopene, and Lutein Content in Tomatoes Grown Under Continuous PAR Spectrum LED Lighting. *Front. Plant Sci.* **2021**, *12*, 611236. [CrossRef]
22. Yousef, A.F.; Ali, M.M.; Rizwan, H.M.; Tadda, S.A.; Kalaji, H.M.; Yang, H.; Ahmed, M.A.A.; Wróbel, J.; Xu, Y.; Chen, F. Photosynthetic apparatus performance of tomato seedlings grown under various combinations of LED illumination. *PLoS ONE* **2021**, *16*, e249373. [CrossRef] [PubMed]
23. Fanasca, S.; Colla, G.; Maiani, G.; Venneria, E.; Rouphael, Y.; Azzini, E.; Saccardo, F. Changes in antioxidant content of tomato fruits in response to cultivar and nutrient solution composition. *J. Agric. Food Chem.* **2006**, *54*, 4319–4325. [CrossRef] [PubMed]
24. Gao, Y.; Yu, H.; Liu, P.; Ma, C.; Li, Q.; Jiang, W. Ending composting during the thermophilic phase improves cultivation substrate properties and increasing winter cucumber yield. *Waste Manag.* **2018**, *79*, 260–272. [CrossRef]
25. Lazzarin, M.; Meisenburg, M.; Meijer, D.; van Ieperen, W.; Marcelis, L.; Kappers, I.; van der Krol, A.; van Loon, J.; Dicke, M. LEDs Make It Resilient: Effects on Plant Growth and Defense. *Trends Plant Sci.* **2021**, *26*, 496–508. [CrossRef] [PubMed]
26. Liu, P.; Li, Q.; Gao, Y.; Wang, H.; Chai, L.; Yu, H.; Jiang, W. A New Perspective on the Effect of UV-B on l-Ascorbic Acid Metabolism in Cucumber Seedlings. *J. Agric. Food Chem.* **2019**, *67*, 4444–4452. [CrossRef] [PubMed]
27. Pan, T.; Wang, Y.; Wang, L.; Ding, J.; Cao, Y.; Qin, G.; Yan, L.; Xi, L.; Zhang, J.; Zou, Z. Increased CO_2 and light intensity regulate growth and leaf gas exchange in tomato. *Physiol. Plant.* **2020**, *168*, 694–708. [CrossRef] [PubMed]
28. Steed, G.; Ramirez, D.C.; Hannah, M.A.; Webb, A.A.R. Chronoculture, harnessing the circadian clock to improve crop yield and sustainability. *Science* **2021**, *372*, 479. [CrossRef] [PubMed]
29. Borchert, R.; Renner, S.S.; Calle, Z.; Navarrete, D.; Tye, A.; Gautier, L.; Spichiger, R.; von Hildebrand, P. Photoperiodic induction of synchronous flowering near the Equator. *Nature* **2005**, *433*, 627–629. [CrossRef]
30. Velez-Ramirez, A.I.; van Ieperen, W.; Vreugdenhil, D.; van Poppel, P.M.J.A.; Heuvelink, E.; Millenaar, F.F. A single locus confers tolerance to continuous light and allows substantial yield increase in tomato. *Nat. Commun.* **2014**, *5*, 4549. [CrossRef]
31. Tewolde, F.T.; Lu, N.; Shiina, K.; Maruo, T.; Takagaki, M.; Kozai, T.; Yamori, W. Nighttime Supplemental LED Inter-lighting Improves Growth and Yield of Single-Truss Tomatoes by Enhancing Photosynthesis in Both Winter and Summer. *Front. Plant Sci.* **2016**, *7*, 448. [CrossRef]
32. Zeng, Z.; Lyu, T.; Jia, X.; Chen, Y.; Lyu, Y. Expression Patterns of Sugar Transporter Genes in the Allocation of Assimilates and Abiotic Stress in Lily. *Int. J. Mol. Sci.* **2022**, *23*, 4319. [CrossRef] [PubMed]
33. Kasperbauer, M.J. Far-Red Light Reflection from Green Leaves and Effects on Phytochrome-Mediated Assimilate Partitioning under Field Conditions. *Plant Physiol.* **1987**, *85*, 350–354. [CrossRef]
34. Wallace, D.H.; Yourstone, K.S.; Masaya, P.N.; Zobel, R.W. Photoperiod gene control over partitioning between reproductive and vegetative growth. *Theor. Appl. Genet.* **1993**, *86*, 6–16. [CrossRef]

35. Lanoue, J.; Zheng, J.; Little, C.; Grodzinski, B.; Hao, X. Continuous Light Does Not Compromise Growth and Yield in Min Cucumber Greenhouse Production with Supplemental LED Light. *Plants* **2021**, *10*, 378. [CrossRef]
36. Nitschke, S.; Cortleven, A.; Iven, T.; Feussner, I.; Havaux, M.; Riefler, M.; Schmülling, T. Circadian Stress Regimes Affect th Circadian Clock and Cause Jasmonic Acid-Dependent Cell Death in Cytokinin-Deficient Arabidopsis Plants. *Plant Cell* **2016**, *2* 1616–1639. [CrossRef]
37. Velez-Ramirez, A.I.; Vreugdenhil, D.; Millenaar, F.F.; van Ieperen, W. Phytochrome A Protects Tomato Plants from Injuries Induce by Continuous Light. *Front. Plant Sci.* **2019**, *10*, 19. [CrossRef] [PubMed]
38. Velez-Ramirez, A.I.; van Ieperen, W.; Vreugdenhil, D.; Millenaar, F.F. Plants under continuous light. *Trends Plant Sci.* **2011**, *1* 310–318. [CrossRef] [PubMed]
39. Ma, D.; Li, X.; Guo, Y.; Chu, J.; Fang, S.; Yan, C.; Noel, J.P.; Liu, H. Cryptochrome 1 interacts with PIF4 to regulate hig temperature-mediated hypocotyl elongation in response to blue light. *Proc. Natl. Acad. Sci. USA* **2016**, *113*, 224–229. [CrossRef]
40. Mao, T.; Li, J.; Wen, Z.; Wu, T.; Wu, C.; Sun, S.; Jiang, B.; Hou, W.; Li, W.; Song, Q.; et al. Association mapping of loci controllin genetic and environmental interaction of soybean flowering time under various photo-thermal conditions. *BMC Genom.* **201** *18*, 415. [CrossRef]

plants

MDPI

Article

Night Interruption with Red and Far-Red Light Optimizes the Phytochemical Composition, Enhances Photosynthetic Efficiency, and Increases Biomass Partitioning in Italian Basil

Soheil Fallah [1], Sasan Aliniaeifard [1,2,*], Mahboobeh Zare Mehrjerdi [1], Shima Mirzaei [1] and Nazim S. Gruda [3,*]

1 Photosynthesis Laboratory, Department of Horticulture, Faculty of Agricultural Technology (Aburaihan), University of Tehran, Tehran 14395-547, Iran; soheilfallah@ut.ac.ir (S.F.); mzarem@ut.ac.ir (M.Z.M.); shimamirzae24@ut.ac.ir (S.M.)

2 Controlled Environment Agriculture Center (CEAC), Faculty of Agricultural Technology (Aburaihan), College of Agriculture and Natural Resources, University of Tehran, Tehran 14395-547, Iran

3 Department of Horticultural Science, INRES-Institute of Crop Science and Resource Conservation, University of Bonn, 53121 Bonn, Germany

* Correspondence: aliniaeifard@ut.ac.ir (S.A.); ngruda@uni-bonn.de (N.S.G.)

Abstract: Controlled environment agriculture is a promising solution to address climate change and resource limitations. Light, the primary energy source driving photosynthesis and regulating plant growth, is critical in optimizing produce quality. However, the impact of specific light spectra during night interruption on improving phytochemical content and produce quality remains underexplored. This study investigated the effects of red (peak wavelength at 660 nm) and far-red night interruption (peak wavelength at 730 nm) on photosynthetic efficiency, biomass distribution, and phytochemical production in Italian basil (*Ocimum basilicum* L.). Treatments included red light, far-red light, a combination of both, and a control without night interruption. Red light significantly increased chlorophyll a by 16.8%, chlorophyll b by 20.6%, and carotenoids by 11%, improving photosynthetic efficiency and nutritional quality. Red light also elevated anthocyanin levels by 15.5%, while far-red light promoted flavonoid production by 43.56%. Although red light enhanced biomass, the primary benefit was improved leaf quality, with more biomass directed to leaves over roots. Far-red light reduced transpiration, enhancing post-harvest water retention and shelf life. These findings demonstrate that red and far-red night interruption can optimize phytochemical content, produce quality, and post-harvest durability, offering valuable insights for controlled environment agriculture. Future research should focus on refining night interruption light strategies across a broader range of crops to enhance produce quality and shelf life in controlled environment agriculture.

Keywords: phytochemical enhancement; light spectrum manipulation; biomass allocation; *Ocimum basilicum*; controlled environment agriculture; photosynthesis

check for updates

Citation: Fallah, S.; Aliniaeifard, S.; Zare Mehrjerdi, M.; Mirzaei, S.; Gruda, N.S. Night Interruption with Red and Far-Red Light Optimizes the Phytochemical Composition, Enhances Photosynthetic Efficiency, and Increases Biomass Partitioning in Italian Basil. *Plants* **2024**, *13*, 3145. https://doi.org/10.3390/plants13223145

Academic Editors: Weijie Jiang and Wenna Zhang

Received: 20 September 2024
Revised: 19 October 2024
Accepted: 4 November 2024
Published: 8 November 2024

1. Introduction

The production of plants under artificial light is a modern agricultural technology that optimizes plant growth independently of seasonal variations in controlled environment agriculture (CEA) systems [1]. In these systems, light-emitting diodes (LEDs) supply the specific light spectra required for plant growth and development [2]. This allows the establishment of ideal environmental conditions, resulting in rapid development, higher yields, and outstanding produce quality while lowering the demand for natural resources and mitigating the impacts of climate change [3,4]. Nevertheless, the expensive initial setup and operational expenses, precise control required over all the environmental cues, and the requirement for specialized expertise and technology have limited the widespread use of this technology for crop production [5].

Providing a proper lighting environment for crops is of immense importance in CEA. Several light attributes, including quality, intensity, and duration, influence the growth and development of crops under CEA conditions. The design and implementation of proper light quality in CEA has attracted attention nowadays. Light quality, including spectral composition and wavelength ratio, strongly affects plant growth, physiology, and development. For instance, red (R) light, with a wavelength between 600 and 700 nm [6], increases growth, photosynthetic capacity, and the production of secondary metabolites that are essential for flavor and food quality [7,8]. Despite some adverse reports related to its sole application, R light enhances plant photosynthesis by increasing the production of photosynthetic pigments, which in turn boosts CO_2 uptake and leads to an increase in biomass [7]. Far-red (FR) light, with a wavelength between 700 and 800 nm, is widely acknowledged as an essential factor affecting plant growth physiology. FR light enhances vegetative growth and stimulates the reproductive development of horticultural crops [6,9]. FR light can either promote or suppress flowering, depending on its presence in the photoperiod [10]. It also contributes to the shade avoidance response. It influences the length of stems and the morphology of plants, resulting in an overall increase in total biomass and growth rate of plants [11]. Plants adjust their growth to compete for light in crowded environments with limited light intensity for normal photosynthesis and plant growth. Phytochromes are R and FR light photoreceptors that regulate plant development through their ability to switch between active (Pfr) and inactive (Pr) forms. This switching mechanism influences seed germination, shade avoidance, flowering, and biomass allocation [12]. The ratio of R to FR light influences the Phytochrome Photo-Stationary State (PSS), affecting the balance between the inactive and active forms of phytochromes (Pr and Pfr forms) [13]. This balance is essential, as it determines how plants respond to their light environment, making these light spectra vital for plant development and adaptation in CEA.

Night interruption (NI) is a horticulture method employed to change the plants' photoperiod by supplying light during the night [14]. This approach subjects plants to night-time light exposure for a specific period, which affects their circadian rhythms and photoperiodic responses [15]. NI can manage energy consumption by changing certain parts of the lighting period to the dark period when demand and expenses are reduced, thus avoiding peak rates while maintaining optimal growth [16]. This method has been reported as a beneficial approach in CEA systems for maximizing plant development and yield [17,18].

Basil (*Ocimum basilicum* L.) is a member of the Lamiaceae family, which includes more than 150 species of aromatic plants. It is native to tropical regions of Asia and Africa [19]. Italian basil is well known for its distinct taste and essential oils. Italian basil cultivars exhibit variations in their physical traits and essential oil composition compared to other sweet basil varieties. They have larger leaves, with high levels of specific metabolites [20,21]. The production of Italian basil under artificial light in CEA has attracted attention. While there were prior studies involving light manipulation in other crops, the present study on night interruption with R and FR light usage in basil and its particular influence on phytochemical production presents new knowledge for research in CEA setups. The findings enable the further understanding of how light spectrum manipulation at night could optimize growth and quality in basil and further improve CEA practices at the commercial level. Therefore, the present study was designed to investigate the impact of NI using R and FR light spectra on the photosynthetic efficiency, biomass production, partitioning, and phytochemical production of Italian basil (*Ocimum basilicum* L.). Using NI with R and FR light spectra is expected to enhance plant growth and phytochemical production in Italian basil under CEA conditions.

2. Materials and Methods

2.1. Plant Material and Growth Conditions

The experiment was conducted at the Controlled Environment Agriculture Center (CEAC), Faculty of Agricultural Technology (Aburaihan), College of Agriculture and Natural Resources, University of Tehran, Iran. The seeds of Italian basil were initially planted in 50-cell trays filled with coir pith and perlite at a ratio of 1:2. The seeds were cultivated in a controlled environment under white light, at 25 ± 2 °C and relative humidity between 60 ± 10%. This ensured an optimal setting for germination and initial growth. Once the seeds had successfully germinated and the seedlings had four leaves, they were transplanted into pots with 15 cm diameters and 10 cm height. The pots were filled with a substrate-containing coir pith and perlite at a ratio of 1:1. This ratio promotes the growth of roots and provides stability for the plants. Subsequently, the plants were irrigated with Hoagland solution [22] by adjusting the pH to 5.7, and the electrical conductivity of the nutrient solution was 1.8 ds m^{-1}.

2.2. Treatments

A mixture of red and blue light spectra was used at a ratio of 70% red light to 30% blue light (R-B), with an intensity of 200 µmol m^{-2} s^{-1} [23], provided by LED light modules (Parcham Company, Pakdasht, Tehran, Iran). The light period was 12 h (6 a.m. to 6 p.m.). This combination of light was used to grow all the plants during the light period. The experimental treatments included different NI lighting spectra, including NI with R light spectrum (NI-R), NI with FR light spectrum (NI-FR), NI with a combination of R and FR light spectra in a 1:1 ratio (NI-R-FR), and a control treatment without NI (C). During the dark period (6 p.m. to 6 a.m.), the NI was applied for two hours, from 12 a.m. to 2 a.m. The intensity of the R and FR light was 100 and 30 µmol m^{-2} s^{-1}, respectively. The intensities for the R-FR were kept at 100 µmol m^{-2} s^{-1} by adjusting the height of the light modules from the plant canopy.

The treatments were continued for 4.5 weeks after applying. Finally, the collected samples were sent to the laboratory for additional analysis.

The daily light integral of 8.64 mol m^{-2} d^{-1} was maintained across all treatments, ensuring consistency in the amount of light each plant received apart from NI light treatments in the experiment. Therefore, 0.72 mol m^{-2} d^{-1} of DLI was added to the NI-R and NI-R-FR. The added DLI for NI-FR was 0.22 mol m^{-2} d^{-1}. Light intensities and spectra were carefully monitored using a Sekonic light meter (Sekonic C-7000, Tokyo, Japan). The spectral composition of different light treatments is presented in Figure 1. PSSs for R-B, NI-R, NI-FR, and NI-R-FR were 0.87, 0.89, 0.14, and 0.87, respectively. The PSS was calculated based on the method described by Sager et al. [24]. The treatments were applied for 4.5 weeks before the collected samples were sent to the laboratory for further analysis. The growth chambers maintained an average temperature of 25 ± 2 °C, a CO_2 concentration of 400 ± 50 ppm, and a relative humidity of 50 ± 5%.

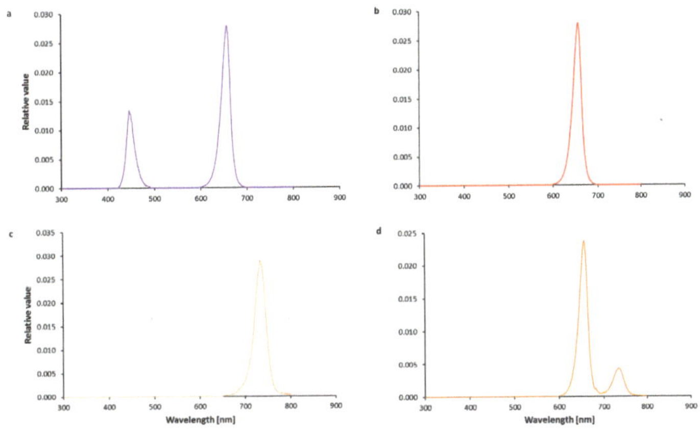

Figure 1. Spectral composition of light used as a control without NI [(**a**); red (peak wavelength at 660 nm) and blue (peak wavelength at 455 nm)], which was used for growing plants during the light period or as night interruption using different light spectra, including red light ((**b**); peak wavelength at 660 nm), far-red light ((**c**); peak wavelength at 630 nm), and a combination of both (**d**).

2.3. *Phytochemical Measurements*

2.3.1. Extraction

Young adult-developed leaves were picked from the plants. Liquid nitrogen wa poured over them, and the samples were powdered using a mortar and pestle. Th powdered samples were used for the extraction process. To measure total phenols an flavonoids, 300 mg of plant sample powder was mixed with 3 mL of 80% methano (1:10 ratio). This mixture was placed in an ultrasonic device at 40 °C for 20 min [25 Subsequently, the samples were centrifuged at 3000× *g* for 15 min, and the supernatan was separated as the extract for evaluating the compounds mentioned above.

150 mg of plant sample powder was mixed with 2 mL of ethanol acidified wit 1% hydrochloric acid to prepare the extract for measuring anthocyanin content. Thi mixture was then placed in a shaker incubator at 4 °C for 24 h. The samples were centrifuge at 5500× *g* for 5 min, and the supernatant was used as the extract.

2.3.2. Measurement of Phenols, Flavonoids, and Anthocyanins

The Folin-Ciocalteu reagent was used to assess the total phenol content by recordin the absorbance of the samples at a wavelength of 730 nm. A calibration curve was draw using gallic acid concentrations of 0, 100, 200, 300, 400, and 500 mg mL^{-1} [26]. The optica absorbance of flavonoids was measured at a wavelength of 415 nm to evaluate the flavonoi content. Different concentrations of quercetin were used to plot the standard curve [27]. T measure the anthocyanins content, the absorbance of the supernatant obtained from th extract in Section 2.3.1 was read at 530 nm and 657 nm [28]. The anthocyanin content wa calculated using the following formula:

Relative Anthocyanin Content = Absorbance at 530 nm − (Absorbance at 657 nm × 0.25

2.4. *Measurement of Photosynthetic Pigments*

To determine the amounts of chlorophyll a, chlorophyll b, and carotenoids, 150 mg of leaf powder was mixed with 2.5 mL of 80% acetone. The mixture was then centrifuged at 8000 rpm for 10 min. The supernatant was collected, and its absorbance was mea sured at wavelengths of 646 nm, 663 nm, and 470 nm. The concentrations of chloro phyll a, chlorophyll b, and carotenoids were calculated using the formulas provided by Lichtenthaler (1987) [29].

2.5. *Post-Harvest Measurements*

Leaf Desiccation Response

Leaf desiccation was used as an early postharvest response of excised leaf to posthar vest conditions. Leaves were excised and re-cut under water. Plant leaves were first saturated for one hour at 21 °C and 35 µmol m^{-2} s^{-1} irradiance, provided by a white LED from Parcham Company, Pakdasht, Tehran, Iran, by placing them in a parafilm-closed container with degassed deionized water. Following saturation, the leaves were posi tioned upside down on balances in a controlled setting with 40 ± 3% relative humidity 21 °C, 1.40 kPa VPD, and 35 µmol m^{-2} s^{-1} irradiance. The weight loss of the leaves was monitored gravimetrically for two hours using an HR200 scale with 0.0001 g accuracy ImageJ v8 (National Institutes of Health, Bethesda, MD, USA) software determined the leaf area in scanned leaves. The transpiration rate was measured by a method described previously [30]. The transpiration rate was calculated using the following formula:

$$TR = \frac{\Delta DW}{\Delta t} \times \frac{1}{M(H_2O)} \times 1000 \times LA$$

where ΔDW is the difference in dry weight during desiccation, Δt is the time difference during measurement, $M(H_2O)$ is the molecular weight of water (18 g mol^{-1}), and LA is the leaf area.

2.6. Biomass Partitioning

Plants were harvested, and the roots, stems, and leaves were separated. The plant components were dried in an oven at 75 °C for 72 h. The dry weights of leaves, roots, and stems were measured using an HR200 scale with 0.0001 g accuracy.

2.7. Photosynthetic Efficiency

Chlorophyll fluorescence was measured using a FluorPen PAR-FLORPEN FP 100-MAX device (Photon System Instrument, Brno, Czech Republic) using the OJIP protocol. The measurements were taken from young, fully developed leaves during week 4 of the vegetative growth stage, after the plants had been dark-adapted for at least 30 min. The dark adaptation was carried out in the middle of the light period. The Performance Index on Absorption Basis (PI_{ABS}), the Dissipation per Reaction Center (DI_O/RC), the minimum and maximum fluorescence yield of the dark-adapted state (F_o and F_m), and the maximum quantum yield of Photosystem II (F_v/F_m) were calculated based on the setting of the device.

2.8. Statistical Analysis

This study used a completely randomized design, with four treatments and three replications (each replication included five plant samples). The data collected were analyzed using the SAS (Statistical Analysis System, version 9.4). Mean comparisons were performed and analyzed using Duncan's multiple-range test. The correlation between traits was determined by calculating the Pearson correlation coefficients (r) using the R package (Version 4.3.3).

3. Results

3.1. Photosynthetic Pigments

Statistical analysis showed significant differences among the studied treatments for the photosynthetic pigments of Italian basil (*Ocimum basilicum* L.) exposed to different light conditions (Table 1).

Table 1. Statistical analysis results of the photosynthetic pigments of the Italian basil (*Ocimum basilicum* L.) exposed to different light treatments.

Sources of Variation	Degrees of Freedom	Mean Square			
		Chlorophyll a	Chlorophyll b	Carotenoid	Total Photosynthetic Pigments
Light Treatment	3	22,580.1 **	3287.17 **	404.76 **	50,380.21 **
Experimental Error	8	254.15	39.46	2.93	387.14

** means significant difference at 1% level.

The analysis of photosynthetic pigments revealed that the NI-R treatment significantly outperformed all other treatments across all four parameters measured (Figure 2). Specifically, chlorophyll content under NI-R reached 530 μg g FW^{-1}, representing a 16.83% increase compared to the control. Similarly, chlorophyll b content was significantly enhanced, with 157.78 μg g FW^{-1} recorded, 20.6% higher than in the control. Carotenoid content also increased, reaching 101.23 μg g FW^{-1}, reflecting a 10.9% rise over the control. The total photosynthetic pigment content under NI-R reached 788.4 μg g FW^{-1}, a 16.7% increase relative to the control chlorophyll content.

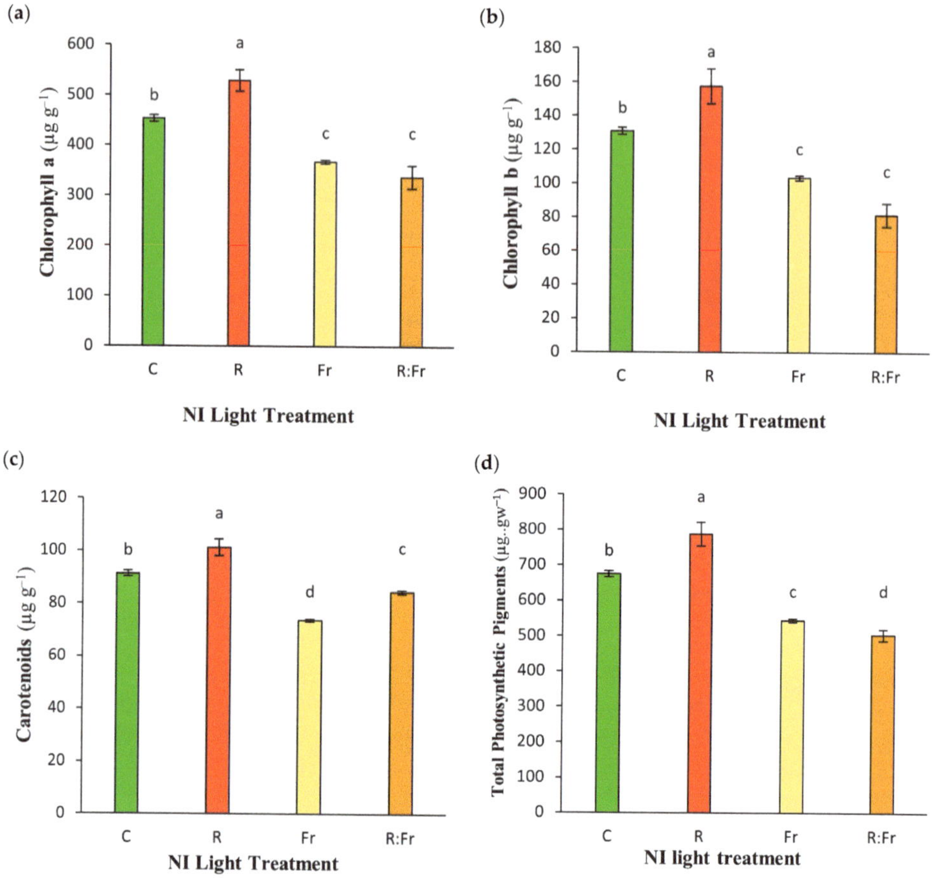

Figure 2. (**a**) Concentration of chlorophyll a; (**b**) chlorophyll b; (**c**) carotenoids; and (**d**) total photosyn thetic pigments (chlorophyll a, b, and carotenoids) of Italian basil plants exposed to night interruptior (NI) using different light spectra, including red light (R), far-red light (FR), a combination of both (R:Fr), and a control without NI (C). During the dark period, the NI treatments were used for 2 h The intensity of the R and FR light was 100 and 30 μmol m^{-2} s^{-1}, respectively. The intensities fo the RFR treatment were kept at 100 μmol m^{-2} s^{-1} by adjusting the height of the light modules from the plant canopy. Columns are the mean value of three replications (each replication included five plants). Means with the same letters within the groups are not significantly different ($p < 0.05$).

3.2. Secondary Metabolites

Statistical analysis showed significant differences among studied treatments for the secondary metabolites of Italian basil exposed to different light conditions (Table 2). The highest anthocyanin content was recorded for NI-R, with 31.02 μg g FW^{-1} (Figure 3a) reflecting a 15.53% increase compared to the control, which recorded 26.85 μg g FW^{-1} Anthocyanin levels in NI-R were higher than in the other light conditions. For NI-R-FR anthocyanin content was measured at 29.19 μg g FW^{-1}, while NI-FR and C were relatively low, with 27.20 μg g FW^{-1} and 26.85 μg g FW^{-1}, respectively. NI-R-FR showed the highest total phenol content, at 0.927 μg g FW^{-1}, which is 109.73% higher than C, with a phenol content of 0.442 μg g FW^{-1} (Figure 3b). The flavonoid content in NI-FR was 1.45 μg g FW^{-1}, which is 43.56% higher than in C (1.01 μg g FW^{-1}). Flavonoid levels in NI-R and NI-R-FR were 1.36 μg g FW^{-1} and 1.33 μg g FW^{-1}, respectively (Figure 3c).

Table 2. Statistical analysis results of the secondary metabolites of the Italian basil (*Ocimum basilicum* L.) exposed to different light treatments.

Sources of Variation	Degrees of Freedom	Mean Square		
		Anthocyanin	Total Phenol	Flavenoids
Light Treatment	3	11.21 **	0.14 **	0.11 **
Experimental Error	8	0.34	0.004	0.001

** means significant difference at 1% level.

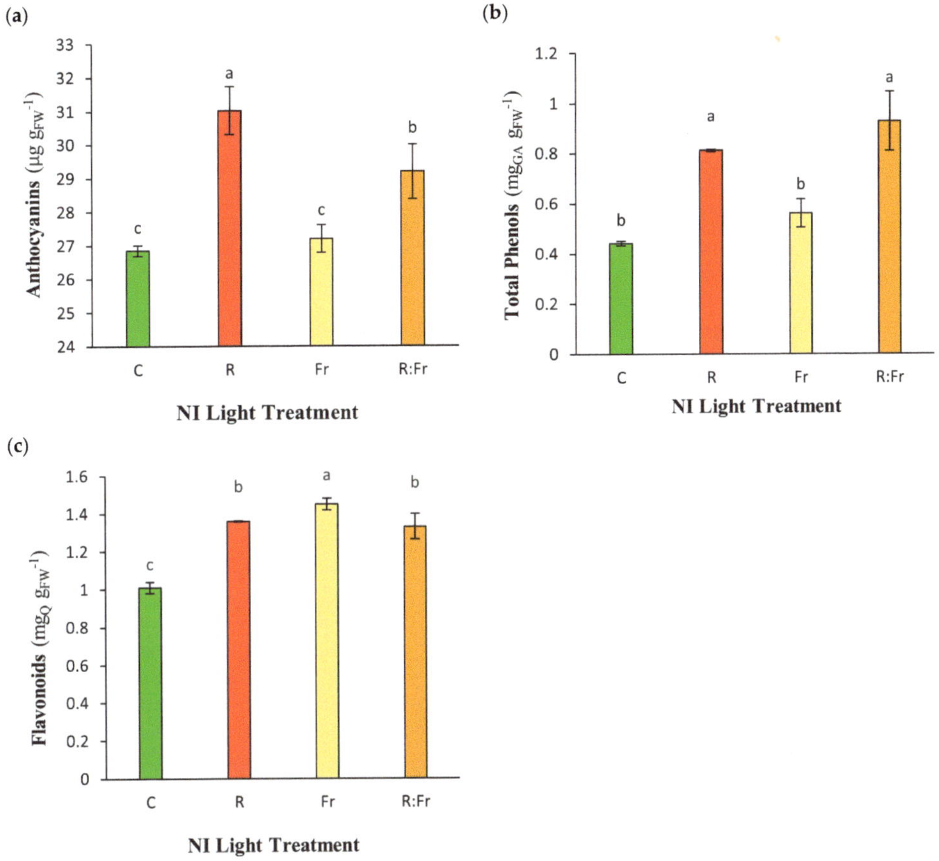

Figure 3. (**a**) Contents of anthocyanins; (**b**) total phenols; and (**c**) flavonoid of Italian basil plants exposed to night interruption (NI) using different light spectra. See Figures 1 and 2 legends for details about light intensities and treatments.

3.3. Leaf Desiccation Response

To study the water loss property of the plants following excision from the plants, the transpiration rate of the excised leaf samples was tracked at the beginning and end of two hours of desiccation. The control samples had the highest transpiration rate, reaching a peak of around 3 mmol m^2 s^{-1} at the beginning of the desiccation. On the other hand, the FR treatment (yellow line) had the lowest transpiration rate, approximately 1.5 mmol m^2 s^{-1}. Throughout the observation period, all treatments consistently decreased transpiration rates with different degrees of slope. The C treatment demonstrated the most significant decline with a slope of −0.0222, indicating the fastest rate of decrease. However, it consistently maintained the highest level, and ended with 0.4 mmol m^2 s^{-1} of transpiration rate. In contrast, the FR light treatment had the most gradual decrease, with a slope of −0.0091 indicating a reduction of approximately 59.01% that was less steep than the C treatment. The R light treatment reached a transpiration rate of 0 by the end of the desiccation period (Figure 4).

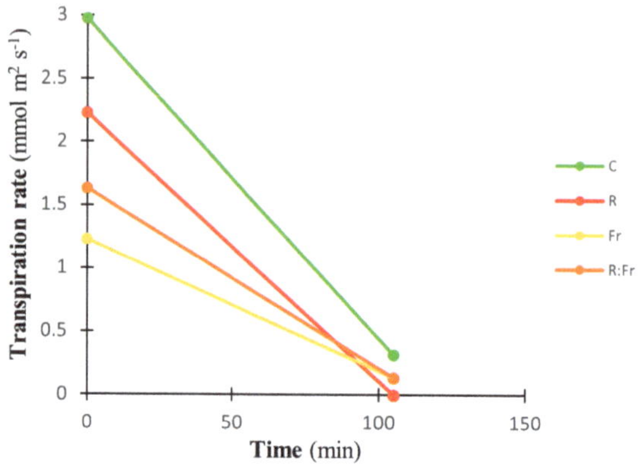

Figure 4. Transpiration rate during two hours of leaf desiccation of Italian basil plants exposed to night interruption (NI) using different light spectra. See Figures 1 and 2 legends for details about light intensities and treatments. The leaves from different treatments were positioned upside down on balances in a controlled setting with 40 ± 3% relative humidity, 21 °C, 1.40 kPa VPD, and 35 µmol m^{-2} s^{-1} irradiance.

3.4. Biomass Production and Partitioning

Statistical analysis showed significant differences among studied treatments for the dry weight of different organs of Italian basil (*Ocimum basilicum* L.) exposed to various light treatments (Table 3). The root, stem, and leaf dry weights demonstrated significant differences under the NI treatments compared to the control (Figure 5). The leaf dry weight showed the most difference, with NI-R-FR reaching 11.785 g, marking a 167.3% increase over the control, which recorded 4.41 g. Stem dry weight followed a similar trend, with NI-R-FR leading at 1.83 g, a 134.6% improvement compared to the control, which was closely matched by NI-R at 1.81 g. For root dry weight, NI-R-FR and NI-FR showed no significant differences, but they differed significantly from the control. NI-R-FR had the highest value at 1.5 g, representing a 92.3% increase compared to the control. The application of NI increased plant height (Figure 5d). The shortest plants were observed under control conditions, while the tallest plants were detected under NI application, including FR light (NI-R-FR and NI-FR).

Table 3. Statistical analysis results of leaf dry weight, stem dry weight, root dry weight of the Italian basil (*Ocimum basilicum* L.) exposed to different light treatments.

Sources of Variation	Degrees of Freedom	Mean Square			
		Leaf Dry Weight	Stem Dry Weight	Root Dry Weight	Plant Height
Light Treatment	3	15.73 **	0.311 **	730.08 **	266 *
Experimental Error	8	0.56	0.01	2.86	4.99

* and ** mean significant difference at 5% and 1% levels, respectively.

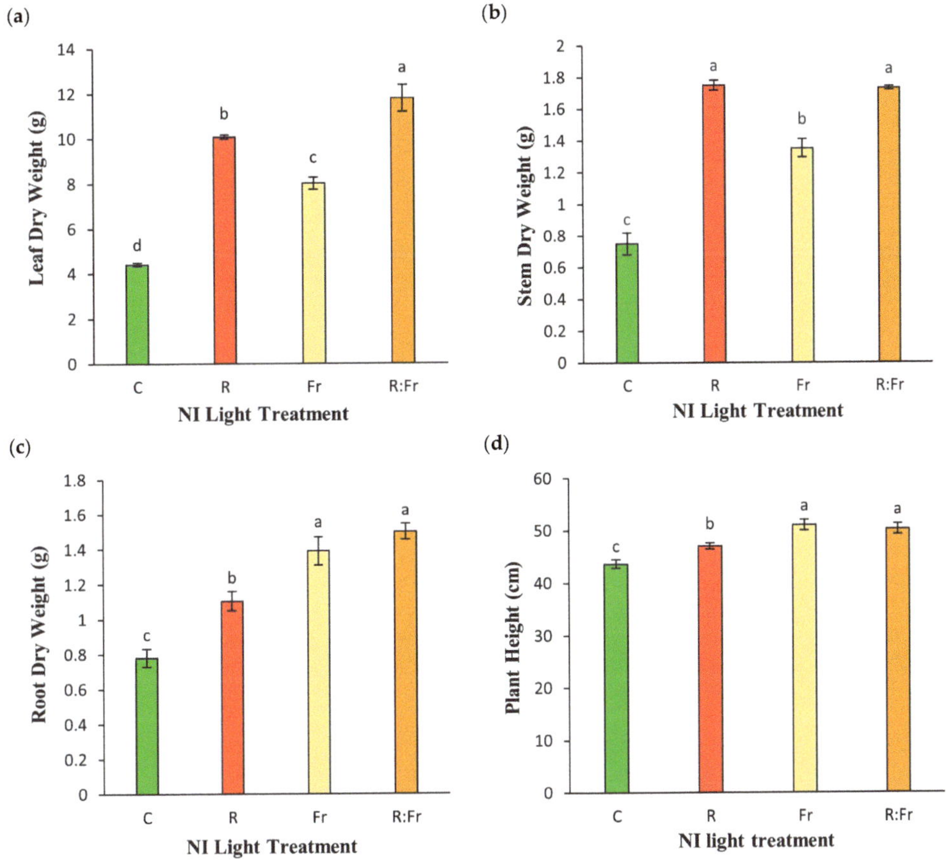

Figure 5. (**a**) Leaf dry weight; (**b**) stem dry weight; (**c**) root dry weight; and (**d**) plant height of Italian basil plants exposed to night interruption (NI) using different light spectra including. See Figures 1 and 2 legends for details about light intensities and treatments. Columns are the mean value of three replications (each replication included five plants). Means with the same letters within the groups are not significantly different ($p < 0.05$).

In general, biomass production in plants exposed to NI-R-FR and NI-R showed more than two times the overall biomass of basil plants compared to the control (Figure 6a). In all treatments, the leaves had the highest biomass allocation. The NI-R-FR treatment had the highest leaf biomass proportion (78.49%), followed by NI-R (77.92%) and NI-FR (74.51%). The biomass contribution to the stem differed significantly among treatments. The NI-R treatment contributed the most stem biomass (13.53%), followed by NI-FR (12.56%) and NI-R-FR (11.52%). Additionally, NI-R contributed the smallest percentage of root biomass among all treatments. In contrast, the control had the highest root biomass proportion (13.13%), while NI-R-FR contributed the least (11.52%) (Figure 6b).

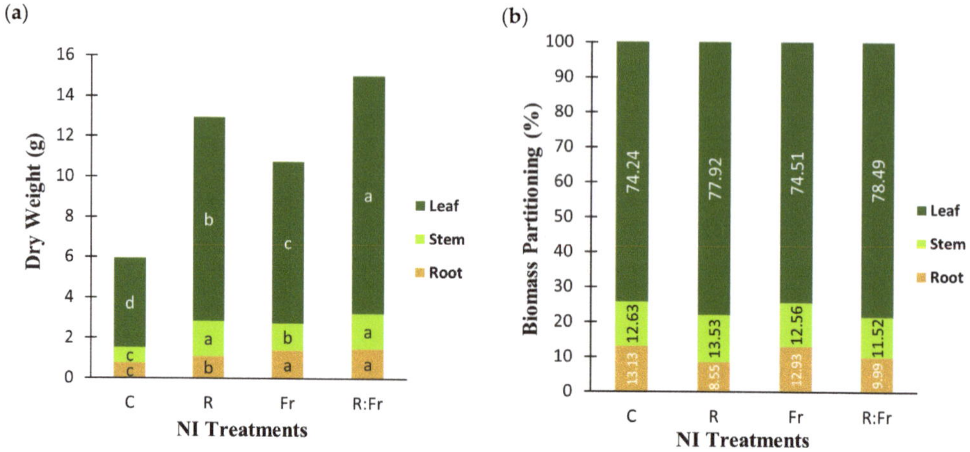

Figure 6. (**a**) Biomass production and (**b**) partitioning of different organs of Italian basil plants exposed to night interruption (NI) using different light spectra. See Figures 1 and 2 legends for details about light intensities and treatments. Columns are the aggregate value of biomass (**a**) or percentage of biomass allocation to each organ (**b**) of plants exposed to different spectra of NI.

3.5. *Photosynthetic Efficiency*

Statistical analysis showed significant differences among studied treatments for some parameters obtained from the OJIP test of Italian basil (*Ocimum basilicum* L.) exposed to different light treatments (Table 4).

Table 4. Statistical analysis results for some of the parameters obtained from the OJIP test of the Italian basil (*Ocimum basilicum* L.) exposed to different light treatments.

Sources of Variation	Degrees of Freedom	Mean Square			
		F_v/F_m	DI_O/RC	PI_{ABS}	F_o
Light Treatment	3	91×10^{-6} ns	0.0026 *	0.089 *	3,455,077.98 *
Experimental Error	8	21×10^{-6}	0.00045	0.019	6,929,404.07

ns and * mean no significant difference and 1% significance level, respectively.

The OJIP test showed varying responses to the NI treatments. No significant difference was detected for F_v/F_m in plants exposed to different NI light (Figure 7a). PI_{ABS} was highest in NI-R and NI-R-FR, measuring 1.7 and 1.75, respectively, with no significant difference. NI-FR, with a PI_{ABS} of 1.6, was not significantly different from NI-R or NI-R-FR, but it was significantly higher than the control with 1.45. For DI_O/RC, the control and NI-FR had similar values with no significant differences. NI-R-FR (0.8) and NI-R (0.78) had significantly lower DI_O/RC than the control, but NI-R-FR was not significantly different from the control or NI-FR. *Fo* was highest in the control at approximately 15,000, while all NI treatments had lower values, around 13,000, indicating a significant decrease compared to the control. F_v/F_m did not vary among treatments.

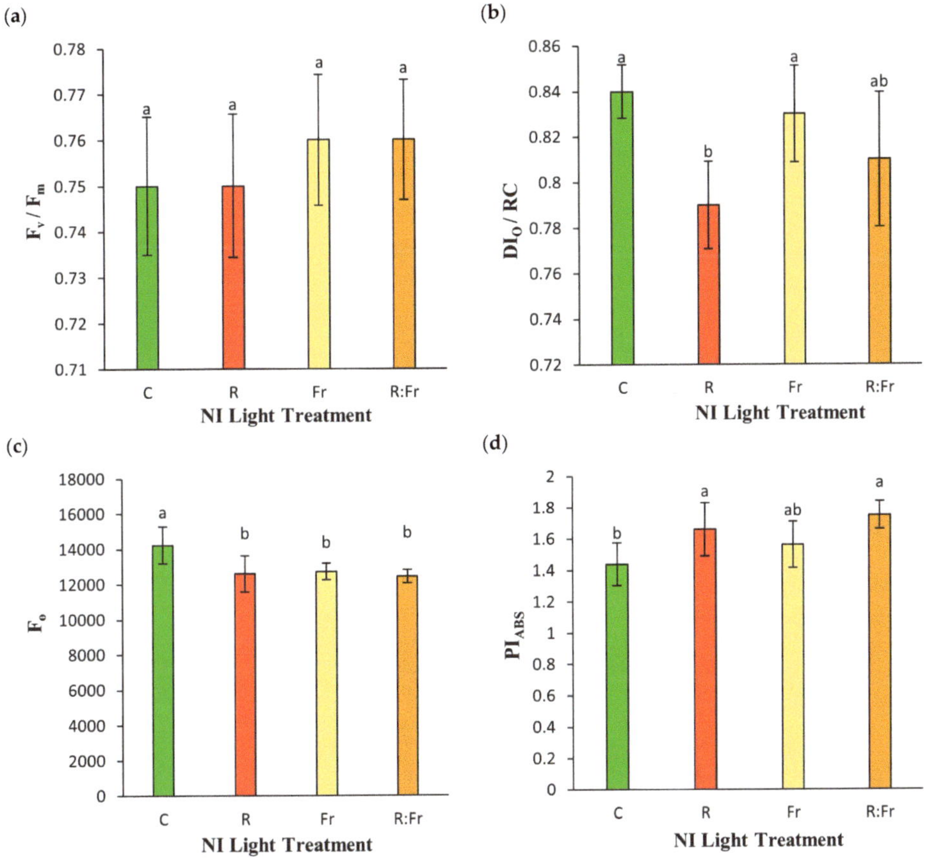

Figure 7. (**a**) Maximum quantum yield of photosystem II (F_v/F_m); (**b**) energy dissipation per reaction center (DI_O/RC); (**c**) minimum fluorescence (F_o); and (**d**) performance index per absorbed light (PI_{ABS}) of Italian basil plants exposed to night interruption (NI) using different light spectra. See Figures 1 and 2 legends for details about light intensities and treatments. Columns are the mean value of three replications (each replication included five plants). Means with the same letters within the groups are not significantly different ($p < 0.05$).

The results of the correlation analysis revealed significant relationships among various biochemical and physiological traits in Italian basil exposed to different light treatments (Table 5). There are high positive correlations between chlorophyll a (1) and chlorophyll b (2) (r = 0.99 **), as well as between chlorophyll b (2) and carotenoids (3) (r = 1.00 **). Anthocyanin content (4) shows strong negative correlations with DI_o/RC (14) (r = 0.99 **). Total phenolic content (5) exhibits negative correlations with antioxidant activity (7) (r = −0.76**) and high positive correlations with leaf dry weight (8) (r = 0.96 *). Flavonoids (6) demonstrate significant positive correlations with leaf area (r = 0.98 *). Strong positive correlations exist among the following dry weight components: leaf dry weight (8), stem dry weight (9), root dry weight (10), and total plant dry weight (11). Notably, leaf dry and plant dry weights have a high correlation (r = 0.99 **). DI_o/RC (14) shows a strong negative correlation with anthocyanin (4) (r = −0.99 **), while PI_{ABS} (15) is positively correlated with total phenolic content (5), and plant dry weight (11) (r = 0.98 *–0.99 **). F_0 demonstrates strong positive correlations with antioxidant activity (7) (r = 0.99 **) and significant negative correlations with stem dry weight (9), plant dry weight (11), and leaf area (12) (r = −0.94 * to −0.97 **).

Table 5. Correlation matrix of biochemical and physiological traits in Italian basil exposed to different light treatments.

	Correlations															
	1	**2**	**3**	**4**	**5**	**6**	**7**	**8**	**9**	**10**	**11**	**12**	**13**	**14**	**15**	**16**
Chlorophyll a (1)	1															
Chlorophyll b (2)	0.99 **	1														
Carotenoids (3)	0.99 **	1.00 **	1													
Anthocyanin (4)	0.46	0.38	0.38	1												
Total phenolic content (5)	−0.13	−0.23	−0.23	0.81	1											
Flavonoids (6)	−0.27	−0.24	−0.24	0.41	0.52	1										
Antioxidant activity (7)	0.40	0.42	0.42	−0.52	−0.76	−0.93	1									
Leaf dry weight (8)	−0.25	−0.32	−0.32	0.73	0.96 *	0.71	−0.90	1								
Stem dry weight (9)	−0.06	−0.12	−0.12	0.83	0.92	0.79	−0.90	0.96 *	1							
Root dry weight (10)	−0.70	−0.71	−0.71	0.22	0.64	0.82	−0.93	0.80	0.72	1						
Plant dry weight (11)	−0.27	−0.34	−0.34	0.72	0.95 *	0.75	−0.92	0.99 **	0.97 *	0.82	1					
Leaf area (12)	−0.35	−0.35	−0.35	0.46	0.64	0.98 *	−0.98	0.81	0.85	0.90	0.84	1				
F_v/F_m (13)	−0.92	−0.90	−0.90	−0.22	0.30	0.62	−0.69	0.48	0.36	0.90	0.51	0.68	1			
DIo/RC (14)	−0.40	−0.32	−0.32	−0.99 **	−0.83	−0.51	0.60	−0.78	−0.88	−0.31	−0.77	−0.55	0.13	1		
PI_{ABS} (15)	−0.26	−0.34	−0.34	0.74	0.98 *	0.63	−0.86	0.99	0.94	0.76	0.99 **	0.75	0.45	−0.78	1	
F_0 (16)	0.27	0.30	0.30	−0.63	−0.81	−0.92	0.99 **	−0.93	−0.95 *	−0.87	−0.94 *	−0.97 *	−0.59	0.71	−0.89	1

*. Correlation is significant at the 0.05 level (2-tailed). **. Correlation is significant at the 0.01 level (2-tailed).

4. Discussion

In the present study, photosynthetic pigment levels were increased by NI-R, while using FR for NI hurt the pigment levels (Figure 2). Chlorophyll a and b are vital pigments that play a critical role in photosynthesis and are effective in light energy absorption. The observed increase in chlorophyll a and b content under NI-R treatment can be attributed to the high efficiency of photosynthesis influenced by R light [31,32]. The R light is recognized as a highly effective light spectrum for enhancing the photosynthetic process because it is absorbed by chlorophyll molecules, leading to increased light energy absorption in an efficient manner [33,34]. It has been reported that R light stimulates the biosynthesis of protochlorophyllide, a precursor in chlorophyll synthesis, thereby increasing chlorophyll content. In contrast, the lower chlorophyll content under NI-FR and NI-R-FR may be due to the inhibitory effect of FR light, which disrupts the biosynthetic pathway of chlorophyll precursors [35]. FR light is less effective in photosynthesis and more effective in morphological modification, which leads to a phenomenon known as the shade avoidance response. Under these conditions, plants prioritize stem growth to escape the shade of other plants, often limiting chlorophyll production [36,37]. It has been shown that R and FR light applications as NIs differentially affect the chlorophyll content of different plant species [38]. Short-day and long-day plants reduce their chlorophyll content when exposed to FR light for four hours during the night (dark) period. Conversely, R light in the context of NI increases the chlorophyll contents compared to FR light-exposed plants [38], which aligns with the findings reported in the present study (Figure 2).

Carotenoids are essential for photoprotection and light harvesting in photosynthesis. The highest carotenoid content was also detected under the NI-R treatment. This response is likely attributed to the role of R light in enhancing carotenoid biosynthesis pathways. Carotenoids act as accessory pigments, absorbing light energy and protecting chlorophyll from oxidative damage [39,40]. The lower carotenoid content under NI-FR may be due to reduced energy absorption and a decreased need for light protection, as FR has less energy than lower wavelengths. FR light significantly impacts the electron density in the photosynthetic reaction centers, particularly affecting Photosystem II (PSII) and Photosystem I (PSI), as it does not provide sufficient energy to effectively drive the water-splitting reaction, leading to fewer electrons entering the electron transport chain and a decrease in photosynthetic efficiency. Although PSI can absorb FR light more effectively, it still suffers from an electron deficiency due to the insufficient electron flow from PSII, resulting in the accumulation of oxidized electron carriers and further reducing the electron density in PSI reaction centers. This imbalance can increase the production of reactive oxygen species (ROS), which can damage cellular components and, if FR light predominates for an extended period, may reduce growth and productivity [41]. On the other hand, it has been shown that FR light reduces the electron density in reaction centers, thereby decreasing photoinhibition and potentially reducing the need for protective systems [42]. However, this should indicate that the intensity of FR light in the present study was much lower than the R light in the NI treatments. Having a higher light intensity for FR light is not considered rational since it would result in the extreme elongation of the plants, which results in lodging (unpublished data). The high positive correlations between chlorophylls and carotenoids (Table 5) indicate that effective light treatment enhances pigment synthesis, which is crucial for photosynthesis and photoprotection.

A balanced light environment can produce secondary metabolites, including antioxidants, at a standardized level, which increases with increased stress [43,44]. The present study's findings indicate that plants under NI treatments did not perform better under extended light exposure as the result of two more hours of light. Since Italian basil is a facultative long-day plant, extending the light through NI increased photosynthetic efficiency and reduced short-light duration, thereby increasing antioxidant properties [45].

Anthocyanins are considered screens during light exposure. They play a role in light protection and the coloration of plant tissues. The highest anthocyanin content under the NI-R treatment may be explained by the activation of the anthocyanin biosynthesis pathway under R light. It has been reported that exposure to R light promotes the expression of genes involved in anthocyanin synthesis, such as chalcone synthase and dihydroflavonol 4-reductase [46,47]. Furthermore, R light also increases the synthesis of phenylalanine ammonia-lyase (PAL), a key enzyme in the phenylpropanoid pathway, leading to increased synthesis of phenolic compounds [45]. Phenolic compounds play a role in the plant's defense system and its response to stress, and their production can be influenced by light quality [48,49]. The results showing the highest flavonoid content under the NI-FR may be due to the plant's response to FR light, which can stimulate flavonoid synthesis as part of the shade avoidance response [40,50]. The lower flavonoid content under the control treatment may be due to a balanced light environment without the additional influence of R or FR light. In a balanced situation, stimulating flavonoid biosynthesis is unnecessary for plant metabolite modification. A negative correlation exists between total phenolic content and antioxidant activity, while a positive correlation exists between total phenolic content and leaf dry weight (Table 5), indicating that phenolics may enhance structural growth without necessarily boosting antioxidant levels. The significant positive correlation between flavonoids and leaf area suggests that larger leaves can enhance flavonoid synthesis, contributing to plant health.

The biomass partitioning observed under various NI treatments highlights the critical role of R and FR light in regulating plant growth, mainly through their influence on the Phytochrome Steady State (PSS). The NI-R-FR treatment resulted in the highest total biomass and the most uniform distribution across plant organs [51]. This outcome can be attributed to the PSS, which reflects the balance between active (Pfr) and inactive (Pr) phytochrome forms. NI affects this balance by regulating the dynamic interconversion between Pr and Pfr. RL drives the conversion of Pr to Pfr, resulting in a higher P_{fr}/P_{Total} ratio, while FRL induces the reverse process, shifting the PSS towards a lower P_{Fr}/P_{Total} ratio [52]. The modulation of the phytochrome system during NI is crucial for certain physiological responses, such as shade avoidance. The increased leaf biomass observed in the NI-FR matches shade avoidance responses, as FRL induces elongation growth in low Pfr conditions [12]. The FR light exposure shifts the PSS towards a lower P_{fr}/P_{Total} ratio inducing shade avoidance responses that promote stem elongation and leaf surface expansion, resulting from the higher leaf biomass in this treatment [53]. R light, which increases the P_{Fr}/P_{Total} ratio, enhances photosynthetic efficiency by stimulating cryptochromes and promoting CO_2 uptake, leading to increased stem and leaf growth and root development by producing growth hormones like auxins [6,54]. The interaction between R light and FR light also influenced electron transport within photosynthetic reaction centers, with R light enhancing and FR light reducing efficiency, contributing to the observed growth patterns [55]. These findings underscore the importance of PSS and light spectra in shaping biomass partitioning, with the NI-R-FR treatment effectively optimizing plant growth and resource allocation [56,57]. However, it is worth noting that increased biomass production resulting from NI application can result from adding more light to the lighting environment. It has been reported that an extended photoperiod is more important than light intensity for biomass gain and phytochemical levels for basil production in controlled environments [58]. However, in the study by Eghbal et al. [58], an hourly increase in the duration of light from 12 to 18 h resulted in less than a 10% increase in biomass. Therefore, in our study, the increase in biomass mainly results from the NI application, and is not due to the extended duration. Future experiments are still needed to elucidate the role of NI in biomass and phytochemical production in basil plants. The same lights should be added to the growing light treatment for the daylight period with the NI light applications. Furthermore, they reported no negative impacts of FR light on the growth of basil plants. In their study, the FR light considerably decreased the leaf area, and the promotive effects of FR light on growth were mainly due to stem elongation. Therefore, in our study, the improved growth due to FR light is a result of its application as the NI.

It has been shown that R and FR light applications as the NI differentially affect the growth and flowering of different plant species [38]. Plants reduce their leaf area when exposed to FR for four hours during the night (dark) period. Conversely, R light used in NI increases the leaf area in different plant species [38], which aligns with the findings reported in the present study. Strong correlations among dry weight components of the present study (Table 5) reflect the interconnected nature of biomass accumulation, particularly the critical role of leaf dry weight in total plant growth.

The analysis of photosynthetic efficiency showed that the NI treatments significantly influenced various aspects of photosynthesis in basil plants. The reduction in F_o across NI treatments suggests enhanced photoprotective mechanisms or increased primary energy absorption efficiency through having more open reaction centers [57,59,60]. The consistent F_v/F_m values between treatments indicate stable maximum quantum efficiency of PSII, suggesting that the overall photochemical conversion potential remains unaffected by R and FR light [61]. The lower DI_O/RC observed in the NI-R and NI-R-FR treatments, compared to the control and NI-FR, indicates reduced energy loss per reaction center under R light conditions, highlighting more efficient energy management within the photosystems [62]. The strong negative correlation between anthocyanin content and DI_O/RC (Table 5) suggests that increased anthocyanins, through providing a screen for extra light entrance, reduce energy dissipation from the photosynthetic apparatus. The higher PI_{ABS} in these treatments shows the enhanced performance of PSII under R light [63]. The positive correlation of PI_{ABS} with total phenolic content and plant dry weight (Table 5) indicates that effective light capture is associated with increased biomass. F0's positive correlation with antioxidant activity and negative correlations with biomass suggest that higher antioxidant levels may be linked to stress responses that inhibit growth.

Exposure to NI-R light decreased the transpiration rate following two hours of desiccation. In comparison, the control showed the highest transpiration rate during two hours of leaf desiccation (Figure 4). Stomata are the primary path for leaf water loss. It has been reported that stomatal size decreases, leading to increased stomatal closing ability due to R light exposure [55]. It has been confirmed that a higher surface area to volume ratio in smaller stomata facilitates stomatal reactions [55]. Generating small stomata and elevated responsiveness to desiccation helps leaves keep internal water and decreases their vulnerability to water deficit and desiccation [55].

5. Conclusions

This study demonstrated that NI treatments significantly affected Italian basil's phytochemical content and post-harvest water loss (*Ocimum basilicum* L.). The NI-R treatment significantly increased carotenoids, chlorophyll a, and chlorophyll b, ultimately increasing photosynthetic efficiency. This treatment also increased the anthocyanin content, showing the role of NI-R light in substantial biochemical processes. The NI-FR treatment was less effective in promoting chlorophyll content, but did stimulate flavonoid production. Regarding post-harvest quality, the desiccation response varied significantly among the NI treatments. The control had the highest transpiration rate, indicative of a higher rate of water loss. In contrast, the NI-FR treatment showed the least transpiration, indicating excellent moisture retention ability and a longer post-harvest life. Results showed that using correct NI treatment strategies with R and FR lights enhanced Italian basil's photosynthetic efficiency and phytochemical production in CEA. From the results, it can be concluded that red and far-red light can improve plant growth and phytochemical yield. However, more research should determine the ideal NI duration, light intensity, and cost-effectiveness under commercial production conditions. Generalizing this research to other crops can improve CEA practices' sustainability and commercial applicability.

Author Contributions: Writing—original draft preparation and formal analysis: S.F. Carrying out the experiment and software: S.F. and S.M. Conceptualization, investigation, resources, project administration, and supervision: S.A. and M.Z.M. Writing—review and editing, visualization, supervision, and validation: N.S.G. and S.A. All authors have read and agreed to the published version of the manuscript.

Funding: This work is based upon research funded by the Iran National Science Foundatior No. 4027100.

Data Availability Statement: Data are contained within the article.

Acknowledgments: The authors are grateful for the support of Pak Rostan Cheshmeh Mihan (Par cham) company, the University of Bonn, and Controlled Environment Agriculture Center (CEAC) University of Tehran for supporting this project.

Conflicts of Interest: The authors declare no conflicts of interest.

References

1. Bantis, F.; Chatzigeorgiou, I.; Sismanis, M.; Ntinas, G.K.; Koukounaras, A. Vegetable Production in PFALs: Control of Micro Environmental Factors, Principal Components and Automated Systems. *Agriculture* **2024**, *14*, 642. [CrossRef]
2. Arcel, M.M.; Lin, X.; Huang, J.; Wu, J.; Zheng, S. The Application of LED Illumination and Intelligent Control in Plant Factory, a New Direction for Modern Agriculture: A Review. *J. Phys. Conf. Ser.* **2021**, *1732*, 012178.
3. Gruda, N.; Bisbis, M.; Tanny, J. Influence of Climate Change on Protected Cultivation: Impacts and Sustainable Adaptatior Strategies—A Review. *J. Clean. Prod.* **2019**, *225*, 481–495. [CrossRef]
4. Gruda, N.; Bisbis, M.; Tanny, J. Impacts of Protected Vegetable Cultivation on Climate Change and Adaptation Strategies fo Cleaner Production—A Review. *J. Clean. Prod.* **2019**, *225*, 324–339. [CrossRef]
5. Cai, W.; Bu, K.; Zha, L.; Zhang, J.; Lai, D.; Bao, H.; Bao, H. Energy Consumption of Plant Factory with Artificial Light: Challenges and Opportunities. *arXiv* **2024**, arXiv:2405.09643.
6. Davarzani, M.; Aliniaeifard, S.; Mehrjerdi, M.Z.; Roozban, M.R.; Saeedi, S.A.; Gruda, N.S. Optimizing Supplemental Light Spectrum Improves Growth and Yield of Cut Roses. *Sci. Rep.* **2023**, *13*, 21381. [CrossRef]
7. Lin, K.H.; Huang, M.Y.; Hsu, M.H. Morphological and Physiological Response in Green and Purple Basil Plants (*Ocimum basilicum* under Different Proportions of Red, Green, and Blue LED Lightings. *Sci. Hortic.* **2021**, *275*, 109677. [CrossRef]
8. Eskandarzade, P.; Mehrjerdi, M.Z.; Gruda, N.S.; Aliniaeifard, S. Phytochemical Compositions and Antioxidant Activity of Greer and Purple Basils Altered by Light Intensity and Harvesting Time. *Heliyon* **2024**, *10*, e30931. [CrossRef]
9. Runkle, E.S.; Heins, R.D. Specific Functions of Red, Far Red, and Blue Light in Flowering and Stem Extension of Long-Day Plants *J. Am. Soc. Hortic. Sci.* **2001**, *126*, 3. [CrossRef]
10. Yamada, A.; Tanigawa, T.; Suyama, T.; Matsuno, T.; Kunitake, T. Red:Far-Red Light Ratio and Far-Red Light Integral Promote or Retard Growth and Flowering in *Eustoma grandiflorum* (Raf.) Shinn. *Sci. Hortic.* **2009**, *120*, 101–106. [CrossRef]
11. Eskandarzade, P.; Zare Mehrjerdi, M.; Didaran, F.; Gruda, N.S.; Aliniaeifard, S. Shading Level and Harvest Time Affect the Photosynthetic and Physiological Properties of Basil Varieties. *Agronomy* **2023**, *13*, 2478. [CrossRef]
12. Lymperopoulos, P.; Msanne, J.; Rabara, R. Phytochrome and Phytohormones: Working in Tandem for Plant Growth and Development. *Front. Plant Sci.* **2018**, *9*, 1037. [CrossRef] [PubMed]
13. Kalaitzoglou, P.; van Ieperen, W.; Harbinson, J.; van der Meer, M.; Martinakos, S.; Weerheim, K.; Nicole, C.C.S.; Marcelis, L.F.M Effects of Continuous or End-of-Day Far-Red Light on Tomato Plant Growth, Morphology, Light Absorption, and Fruit Production *Front. Plant Sci.* **2019**, *10*, 322. [CrossRef] [PubMed]
14. Bennie, J.; Davies, T.W.; Cruse, D.; Gaston, K.J. Ecological Effects of Artificial Light at Night on Wild Plants. *J. Ecol.* **2016**, *104* 611–620. [CrossRef]
15. Wei, Y.; Li, Z.; Zhang, J.; Hu, D. Influence of Night-Time Light Pollution on the Photosynthesis and Physiological Characteristics of the Urban Plants *Euonymus japonicus* and *Rosa hybrida*. *Ecol. Process* **2023**, *12*, 38. [CrossRef]
16. Ma, L.; Li, Y.; Li, X.; Xu, D.; Lin, X.; Liu, M.; Li, G.; Qin, X. FAR-RED ELONGATED HYPOCOTYLS3 Negatively Regulates Shade Avoidance Responses in Arabidopsis. *Plant Cell Environ.* **2019**, *42*, 3280–3292. [CrossRef]
17. Adams, S.R.; Langton, F.A. Photoperiod and Plant Growth: A Review. *J. Hortic. Sci. Biotechnol.* **2005**, *80*, 2–10. [CrossRef]
18. Frąszczak, B.; Kałużewicz, A.; Krzesiński, W.; Lisiecka, J.; Spiżewski, T. Effect of Differential Temperature and Photoperiod on Growth of *Ocimum basilicum*. *Agriculture* **2011**, *98*, 375–382.
19. Darrah, H.H. Investigations of the Cultivars of Basils (Ocimum). *Econ. Bot.* **1974**, *28*, 63–67. [CrossRef]
20. Kwee, E.M.; Niemeyer, E.D. Variations in Phenolic Composition and Antioxidant Properties among 15 Basil (*Ocimum basilicum* L.) Cultivars. *Food Chem.* **2011**, *128*, 1044–1050. [CrossRef]
21. Marotti, M.; Piccaglia, R.; Giovanelli, E. Differences in Essential Oil Composition of Basil (*Ocimum basilicum* L.) Italian Cultivars Related to Morphological Characteristics. *J. Agric. Food Chem.* **1996**, *44*, 3926–3929. [CrossRef]
22. Hosseini, H.; Mozafari, V.; Roosta, H.R.; Shirani, H.; van de Vlasakker, P.C.H.; Farhangi, M. Nutrient Use in Vertical Farming Optimal Electrical Conductivity of Nutrient Solution for Growth of Lettuce and Basil in Hydroponic Cultivation. *Horticulturae* **2021**, *7*, 283. [CrossRef]
23. Hosseini, A.; Zare Mehrjerdi, M.; Aliniaeifard, S.; Seif, M. Photosynthetic and Growth Responses of Green and Purple Basil Plants under Different Spectral Compositions. *Physiol. Mol. Biol. Plants* **2019**, *25*, 741–752. [CrossRef] [PubMed]
24. Sager, J.C.; Smith, W.O.; Edwards, J.L.; Cyr, K.L. Cyr Photosynthetic Efficiency and Phytochrome Photoequilibria Determination Using Spectral Data. *Trans. ASAE* **1988**, *31*, 1882–1889. [CrossRef]

25. Michalaki, A.; Karantonis, H.C.; Kritikou, A.S.; Thomaidis, N.S.; Dasenaki, M.E. Ultrasound-Assisted Extraction of Total Phenolic Compounds and Antioxidant Activity Evaluation from Oregano (*Origanum vulgare* Ssp. Hirtum) Using Response Surface Methodology and Identification of Specific Phenolic Compounds with HPLC-PDA and Q-TOF-MS/MS. *Molecules* **2023**, *28*, 2033. [CrossRef]
26. Cheng, A.; Chen, X.; Jin, Q.; Wang, W.; Shi, J.; Liu, Y. Comparison of Phenolic Content and Antioxidant Capacity of Red and Yellow Onions. *Czech J. Food Sci.* **2013**, *31*, 501–508. [CrossRef]
27. Javanmardi, J.; Stushnoff, C.; Locke, E.; Vivanco, J.M. Antioxidant Activity and Total Phenolic Content of Iranian Ocimum Accessions. *Food Chem.* **2003**, *83*, 547–550. [CrossRef]
28. Teng, S.; Keurentjes, J.; Bentsink, L.; Koornneef, M.; Smeekens, S. Sucrose-Specific Induction of Anthocyanin Biosynthesis in Arabidopsis Requires the MYB75/PAP1 Gene. *Plant Physiol.* **2005**, *139*, 1840–1852. [CrossRef]
29. Lichtenthaler, H.K. Chlorophylls and Carotenoids: Pigments of Photosynthetic Biomembranes. *Methods Enzym.* **1987**, *148*, 350–382. [CrossRef]
30. Aliniaeifard, S.; Van Meeteren, U. Natural Variation in Stomatal Response to Closing Stimuli among *Arabidopsis thaliana* Accessions after Exposure to Low VPD as a Tool to Recognize the Mechanism of Disturbed Stomatal Functioning. *J. Exp. Bot.* **2014**, *65*, 6529–6542. [CrossRef]
31. Heo, J.W.; Lee, C.W.; Murthy, H.N.; Paek, K.Y. Growth Responses of Lettuce Seedlings under Various Ratios of Red to Far-Red Light-Emitting Diodes. *Hortic. Environ. Biotechnol.* **2012**, *53*, 142–149.
32. Kitajima, K.; Hogan, K. Increases of Chlorophyll a/b Ratios during Acclimation of Tropical Woody Seedlings to Nitrogen Limitation and High Light. *Plant Cell Environ.* **2003**, *26*, 857–865. [CrossRef] [PubMed]
33. Hopkins, W.G. *Introduction to Plant Physiology*; Wiley: Hoboken, NJ, USA, 1999.
34. Jordan, B.R.; He, J.; Chow, W.S.; Anderson, J.M. Increased Sensitivity of Chlorophyll a Fluorescence Parameters to Photoinhibition Brought about by the Susceptibility of Photosystem II to Low-Temperature Photoinhibition. *Plant Cell Environ.* **2001**, *24*, 651–662. [CrossRef]
35. Price, L.; Klein, W.H. Red, Far-Red Response & Chlorophyll Synthesis. *Plant Physiol.* **1961**, *36*, 733–735. [CrossRef]
36. Li, Q.; Kubota, C. Effects of Supplemental Light Quality on Growth and Phytochemicals of Baby Leaf Lettuce. *Environ. Exp. Bot.* **2009**, *67*, 59–64. [CrossRef]
37. Ohashi-Kaneko, K.; Takase, M.; Kon, N.; Fujiwara, K.; Kurata, K. Effect of Light Quality on Growth and Vegetable Quality in Leaf Lettuce, Spinach and Komatsuna. *Environ. Control. Biol.* **2007**, *45*, 189–198. [CrossRef]
38. Kozai, T. Smart Plant Factory. In *The Next Generation Indoor Vertical Farms*; Springer: Singapore, 2018.
39. Havaux, M.; Kloppstech, K. The Protective Functions of Carotenoid and Flavonoid Pigments against Excess Visible Radiation at Chilling Temperature Investigated in Arabidopsis NPQ and TT Mutants. *Planta* **2001**, *213*, 953–966. [CrossRef]
40. Hogewoning, S.W.; Trouwborst, G.; Maljaars, H.; Poorter, H.; van Ieperen, W.; Harbinson, J. Blue Light Dose-Responses of Leaf Photosynthesis, Morphology, and Chemical Composition of Cucumis Sativus Grown under Different Combinations of Red and Blue Light. *J. Exp. Bot.* **2010**, *61*, 3107–3117. [CrossRef]
41. Shomali, A.; De Diego, N.; Zhou, R.; Abdelhakim, L.; Vrobel, O.; Tarkowski, P.; Aliniaeifard, S.; Kamrani, Y.Y.; Ji, Y.; Ottosen, C.-O. The Crosstalk of Far-Red Energy and Signaling Defines the Regulation of Photosynthesis, Growth, and Flowering in Tomatoes. *Plant Physiol. Biochem.* **2024**, *208*, 108458. [CrossRef]
42. Shomali, A.; Ottosen, C.O.; Zhou, R.; Aliniaeifard, S.; Abdelhakim, L. Far-Red Signaling Affects the Regulation of Photosynthesis in Tomatoes. *Acta Hortic.* **2024**, *1391*, 191–196. [CrossRef]
43. Kiferle, C.; Lucchesini, M.; Mensuali-Sodi, A.; Maggini, R.; Raffaelli, A.; Pardossi, A. Rosmarinic Acid Content in Basil Plants Grown in Vitro and in Hydroponics. *Cent. Eur. J. Biol.* **2011**, *6*, 946–957. [CrossRef]
44. Shiga, T.; Shoji, K.; Shimada, H.; Hashida, S.-I.; Goto, F.; Yoshihara, T. Effect of Light Quality on Rosmarinic Acid Content and Antioxidant Activity of Sweet Basil, *Ocimum basilicum* L. *Plant Biotechnol.* **2009**, *26*, 255–259. [CrossRef]
45. Khalighi, S.; Mehrjerdi, Z.; Aliniaeifard, S. Light quality affects phytochemicals content and antioxidant capacity of *Satureja hortensis*. *S. West. J.* **2021**, *12*, 99–117.
46. Stutte, G.W.; Edney, S.; Skerritt, T. Photoregulation of Bioprotectant Content of Red Leaf Lettuce with Light-Emitting Diodes. *HortScience* **2009**, *44*, 79–82. [CrossRef]
47. Wang, Q.; Liu, Q.; Wang, X.; Zuo, Z.; Oka, Y.; Lin, C. New Insights into the Mechanisms of Phytochrome-Cryptochrome Coaction. *New Phytol.* **2018**, *217*, 547–551. [CrossRef]
48. Oliveira, G.C.; Vieira, W.L.; Bertolli, S.C.; Pacheco, A.C. Photosynthetic Behavior, Growth and Essential Oil Production of *Mellisa officinalis* L. Cultivated under Colored Shade Nets. *Chil. J. Agric. Res.* **2016**, *76*, 123–128. [CrossRef]
49. Verma, R.S.; Padalia, R.C.; Chauhan, A. Variation in the Volatile Terpenoids of Two Industrially Important Basil (*Ocimum basilicum* L.) Varieties. *J. Sci. Food Agric.* **2012**, *92*, 626–631. [CrossRef]
50. Vaštakaitė, V.; Viršilė, A.; Brazaitytė, A.; Samuolienė, G.; Jankauskienė, J.; Sirtautas, R.; Novičkovas, A.; Dabašinskas, L.; Sakalauskienė, S.; Miliauskienė, J.; et al. The Effect of Blue Light Dosage on Growth and Antioxidant Properties of Microgreens. *Sodinink. Darzinink.* **2015**, *34*, 25–35.
51. Lobiuc, A.; Vasilache, V.; Pintilie, O.; Stoleru, T.; Burducea, M.; Oroian, M.; Zamfirache, M.M. Blue and Red LED Illumination Improves Growth and Bioactive Compounds Contents in Acyanic and Cyanic *Ocimum basilicum* L. Microgreens. *Molecules* **2017**, *22*, 2111. [CrossRef]

52. Tripathi, S.; Hoang, Q.T.N.; Han, Y.J.; Kim, J. Il Regulation of Photomorphogenic Development by Plant Phytochromes *Int. J. Mol. Sci.* **2019**, *20*, 6165. [CrossRef]
53. Carvalho, S.D.; Schwieterman, M.L.; Abrahan, C.E.; Colquhoun, T.A.; Folta, K.M. Light Quality Dependent Changes in Morphology, Antioxidant Capacity, and Volatile Production in Sweet Basil (*Ocimum basilicum*). *Front. Plant Sci.* **2016**, *7*, 1328 [CrossRef]
54. Bayat, L.; Arab, M.; Aliniaeifard, S.; Seif, M.; Lastochkina, O.; Li, T. Effects of Growth under Different Light Spectra on the Subsequent High Light Tolerance in Rose Plants. *AoB Plants* **2018**, *10*, ply052. [CrossRef] [PubMed]
55. Seif, M.; Aliniaeifard, S.; Arab, M.; Mehrjerdi, M.Z.; Shomali, A.; Fanourakis, D.; Li, T.; Woltering, E. Monochromatic Red Light during Plant Growth Decreases the Size and Improves the Functionality of Stomata in Chrysanthemum. *Funct. Plant Biol.* **2021** *48*, 515. [CrossRef] [PubMed]
56. Matysiak, B.; Kowalski, A. White, Blue and Red LED Lighting on Growth, Morphology and Accumulation of Flavonoid Compounds in Leafy Greens. *Zemdirbyste* **2019**, *106*, 281–286. [CrossRef]
57. Zheng, L.; Van Labeke, M.C. Long-Term Effects of Red- and Blue-Light Emitting Diodes on Leaf Anatomy and Photosynthetic Efficiency of Three Ornamental Pot Plants. *Front. Plant Sci.* **2017**, *8*, 917. [CrossRef]
58. Eghbal, E.; Aliniaeifard, S.; Mehrjerdi, M.Z.; Abdi, S.; Hassani, S.B.; Rassaie, T.; Gruda, N.S. Growth, phytochemical, and phytohormonal responses of basil to different light durations and intensities under constant daily light integral. *BMC Plant Biol.* **2024**, *24*, 935. [CrossRef]
59. Kalaji, H.M.; Jajoo, A.; Oukarroum, A.; Brestic, M.; Zivcak, M.; Samborska, I.A.; Cetner, M.D.; Łukasik, I.; Goltsev, V.; Ladle, R.J. Chlorophyll a Fluorescence as a Tool to Monitor Physiological Status of Plants under Abiotic Stress Conditions. *Acta Physiol. Plant* **2014**, *36*, 102. [CrossRef]
60. Murchie, E.H.; Lawson, T. Chlorophyll Fluorescence Analysis: A Guide to Good Practice and Understanding Some New Applications. *J. Exp. Bot.* **2013**, *64*, 3983–3998. [CrossRef]
61. Strasser, R.J.; Tsimilli-Michael, M.; Srivastava, A. Analysis of the Chlorophyll a Fluorescence Transient. In *Chlorophyll a Fluorescence: A Signature of Photosynthesis*; Papageorgiou, G.C., Govindjee, Eds.; Springer: Dordrecht, The Netherlands, 2004; pp. 321–362.
62. Terashima, I.; Evans, J.R. Effects of Light and Nitrogen Nutrition on the Organization of the Photosynthetic Apparatus in Spinach *Plant Cell Physiol.* **1988**, *29*, 143–155.
63. Li, Y.; Xin, G.; Liu, C.; Shi, Q.; Yang, F.; Wei, M. Effects of Red and Blue Light on Leaf Anatomy, CO_2 Assimilation and the Photosynthetic Electron Transport Capacity of Sweet Pepper (*Capsicum annuum* L.) Seedlings. *BMC Plant Biol.* **2020**, *20*, 318 [CrossRef]

plants

MDPI

Communication

Optimal Planting Time for Summer Tomatoes (*Lycopersicon esculentum* Mill.) Cropping in Korea: Growth, Yield, and Photosynthetic Efficiency in a Semi-Closed Greenhouse

Hyo Jun Bae [1,2,†], Seong-Hoon Kim [3,†], Yuseok Jeong [1], Sungjin Park [1], Kingsley Ochar [3], Youngsin Hong [1], Yun Am Seo [4], Baul Ko [2], Jeong Hyang Bae [2], Dong Soo Lee [5] and Inchan Choi [1,*]

1 Division of Agricultural Engineering, National Institute of Agricultural Sciences, Rural Development Administration, Jeonju 54875, Republic of Korea; bhj0711@korea.kr (H.J.B.); archha98@korea.kr (Y.J.); dbtjr2665@korea.kr (S.P.); honge159@korea.kr (Y.H.)
2 Department of Horticulture Industry, Wonkwang University, Iksan 54538, Republic of Korea; dogwal@hanmail.net (B.K.); bae@wku.ac.kr (J.H.B.)
3 National Agrodiversity Center, National Institute of Agricultural Sciences, Rural Development Administration, Jeonju 54875, Republic of Korea; shkim0819@korea.kr (S.-H.K.); ocharking@korea.kr (K.O.)
4 Department of Data Science, Jeju National University, Jeju 63243, Republic of Korea; seoya@jejunu.ac.kr
5 National Institute of Horticultural & Herbal Science, Wanju 55365, Republic of Korea; tara0808@korea.kr
* Correspondence: inchchoi@korea.kr; Tel.: +82-63-238-4035
† These authors contributed equally to this work.

Abstract: In Korea, greenhouses are traditionally used for crop cultivation in the winter. However, due to diverse consumer demands, climate change, and advancements in agricultural technology, more farms are aiming for year-round production. Nonetheless, summer cropping poses challenges such as high temperatures, humidity from the monsoon season, and low light conditions, which make it difficult to grow crops. Therefore, this study aimed to determine the best planting time for summer tomato cultivation in a Korean semi-closed greenhouse that can be both air-conditioned and heated. The experiment was conducted in the Advanced Digital Greenhouse, built by the National Institute of Agricultural Sciences. The tomato seedlings were planted in April, May, and June 2022. Growth parameters such as stem diameter, flowering position, stem growth rate, and leaf shape index were measured, and harvesting was carried out once or twice weekly per treatment from 65 days to 265 days after planting. The light use efficiency and yield per unit area at each planting time was measured. Tomatoes planted in April showed a maximum of 42.9% higher light use efficiency for fruit production and a maximum of 33.3% higher yield. Furthermore, the growth form of the crops was closest to the reproductive growth type. Therefore, among April, May, and June, April is considered the most suitable planting time for summer cultivation, which is expected to contribute to reducing labor costs due to decreased workload and increasing farm income through increased yields. Future research should explore optimizing greenhouse microclimates and developing crop varieties tailored for summer cultivation to further enhance productivity and sustainability in year-round agricultural practices.

Keywords: cooling; light use efficiency; semi-closed greenhouse; summer cropping; tomato

Citation: Bae, H.J.; Kim, S.-H.; Jeong, Y.; Park, S.; Ochar, K.; Hong, Y.; Seo, Y.A.; Ko, B.; Bae, J.H.; Lee, D.S.; et al. Optimal Planting Time for Summer Tomatoes (*Lycopersicon esculentum* Mill.) Cropping in Korea: Growth, Yield, and Photosynthetic Efficiency in a Semi-Closed Greenhouse. *Plants* **2024**, *13*, 2116. https://doi.org/10.3390/plants13152116

Academic Editors: Weijie Jiang and Wenna Zhang

Received: 21 June 2024
Revised: 16 July 2024
Accepted: 22 July 2024
Published: 30 July 2024

1. Introduction

In Korea, where distinct seasons—spring, summer, autumn, and winter—prevail, greenhouses serve a crucial role in agriculture, particularly for winter cultivation. They provide a warm environment essential for growing crops during cold weather. However, with changing consumer demands and challenges posed by climate change, there is a growing trend towards urban agriculture and year-round production [1]. Climate change has extended summers and introduced abnormal weather patterns, increasing demand for crops during high-temperature periods [2]. Consequently, the area dedicated to facilities for year-round cultivation is expanding annually. Crops experience reduced net photosynthesis due to excessive respiration from high temperatures and intense light, leading to poor growth caused by transpiration and stress [3]. This results in a decline in product quality and yield, necessitating active measures such as shading or cooling to mitigate these effects [4,5]. Furthermore, the current state of agriculture in Korea is characterized by a declining rural population and aging [6], leading to increased labor and management costs and worsening conditions for agricultural production. To address these issues and achieve sustainable agriculture, smart farming has emerged and is being promoted [7]. For smart agriculture, smart farm technologies are being developed for open fields, facility horticulture, and livestock farming [8]. In facility horticulture, various sensors, actuators and environmental control systems are installed in greenhouses and connected to the internet [9]. This allows for the observation of internal conditions and the adjustment of settings to control the internal environment from anywhere, at any time. Additionally external weather stations, internal temperature and humidity sensors, nutrient solution monitors, leaf temperature sensors, and thermal imaging cameras collect environmental data from both inside and outside, providing objective data that increases the convenience of farm operations [10]. The accumulated data analysis enables annual improvements to the internal environment, potentially leading to increased yields.

Greenhouses are categorized into open, semi-closed, and closed types [11–13]. Open greenhouses often have side vents and are commonly single-span structures, while closed greenhouses are vertical indoor farms. A semi-closed greenhouse has a higher internal pressure compared to the outside and minimizes the intake of external air by using active heating and cooling systems [14]. By creating a positive pressure inside the greenhouse the environment can be maintained uniformly and stably, reducing the need for pesticides by preventing pest infiltration [12]. Minimal ventilation with external air saves energy and maintains a high concentration of CO_2, which can increase crop productivity [14–17]. However, currently, more than 95% of Korean greenhouses are plastic-covered, and most of the semi-closed greenhouses in Korea are glass greenhouses. Typically, tomatoes are transplanted in August and grown until June of the following year, allowing for the harvest of more than 30 flower clusters. There have been studies on seedling methods for high-temperature periods [18], optimal pruning management for high-temperature periods [19], and comparisons of paprika growth and yield in semi-closed greenhouses and regular plastic greenhouses during high-temperature periods [20]. However, research on the appropriate transplanting time for summer cropping is still insufficient. In Korea summer cropping is challenging due to high temperatures, high humidity from monsoons and low light conditions, leading to poor fruit set and the occurrence of misshapen fruits which reduces productivity and makes it less favorable for farmers. Nevertheless, domestic tomato production decreases during this period. Therefore, in greenhouses equipped with cooling facilities, an increase in average prices can be expected, potentially boosting farmers incomes. This study aims to determine the most suitable transplanting time for summer cropping of tomatoes in April, May, or June in Korean-style semi-closed plastic greenhouses with heating and cooling capabilities. The current work investigated the growth dynamics yield variations, and chlorophyll fluorescence responses of summer tomatoes planted at different times, aiming to provide insights into optimal strategies of tomato cultivation in summer seasons in Korea. Using an advanced greenhouse facility provides an ideal setting for investigating the growth and development of tomato plants under controlled environmental conditions as well as helping to provide reliable and reproducible data in crop experiments [21].

2. Results

2.1. Growth Analysis Based on the Transplanting Time

Significant differences ($p \leq 0.05$) were observed in stem diameter and flowering position across different planting times. Generally, tomatoes planted in June had the largest growth, while those planted in April had the smallest (Tables 1 and 2).

Table 1. Plant length and leaf shape index of tomato plants according to the planting season.

	Planting Season	Days after Planting (DAP)					
		Early Growth			Late Growth		
		75	82	89	244	251	258
Stem Growth z (cm/week)	April	17.1 c	21.6 b	24.7 a	10.6 c	9.4 c	12.5 c
	May	23.9 a	28.3 a	25.8 a	15.5 b	15.3 b	15.0 b
	June	21.8 b	21.5 b	25.8 a	18.6 a	20.1 a	21.5 a
	LSD (5%) y	***	**	ns	***	***	***
Leaf shape index z (%)	April	1.16 a	1.14 a	1.22 a	1.22 a	1.23 a	1.29 a
	May	1.19 a	1.09 ab	1.08 b	1.07 b	1.01 c	0.96 c
	June	0.98 b	1.01 b	0.96 c	1.05 b	1.15 b	1.08 b
	LSD (5%) y	**	*	***	**	***	***

z The stem growth was measured by determining the distance from the guide string marked at the position of the growing point (epical meristem) 7 days prior to the current growing point. Leaf shape index = leaf length/leaf width. y Mean separation within columns by Duncan's multiple range test at $p = 0.05$. *, **, and ***: Significant at $p \leq 0.05$, 0.01, and 0.001, respectively. The letters a, b, and c indicate the presence or absence of significant variability. The same letters shared between groups signify no significant differences, while different letters indicate significant variability ($p < 0.05$) among the groups.

Table 2. Stem diameter and flowering position of tomato plants according to the planting season.

	Planting Season	Days after Planting (DAP)					
		Early Growth			Late Growth		
		75	82	89	244	251	258
Stem diameter (mm) z	April	9.0 c	8.9 b	9.1 b	10.2 b	8.7 b	9.3 b
	May	11.6 b	9.5 b	11.9 a	12.6 a	11.5 a	8.1 b
	June	13.4 a	12.8 a	12.0 a	13.1 a	11.0 a	12.0 a
	LSD (5%) y	***	**	**	**	**	**
Flowering position z (cm)	April	14.4 c	16.8 c	15.6 b	21.3 b	15.1 c	16.1 b
	May	24.6 b	22.1 b	27.6 a	26.5 a	30.6 a	34.1 a
	June	30.5 a	30.3 a	28.0 a	19.9 b	19.9 b	18.8 b
	LSD (5%) y	***	***	**	**	***	**

z The flowering position was expressed as the distance between the flowering truss and the head of the plant. The stem diameter was measured just below the topmost flowering cluster. y Mean separation within columns by Duncan's multiple range test at $p = 0.05$. **, and ***: significant at $p \leq 0.01$, and 0.001, respectively. The letters a, b, and c indicate the presence or absence of significant variability. The same letters shared between groups signify no significant differences, while different letters indicate significant variability ($p < 0.05$) among the groups.

2.1.1. Stem Growth

The early growth investigation was conducted from 75 to 89 days after transplanting. The stem growth for tomatoes planted in April ranged from 17.1 cm to 24.7 cm, showing no significant differences from other planting times. The late growth investigation was conducted from 244 to 258 days after transplanting. The stem growth for tomatoes planted in April ranged from 9.4 cm to 12.5 cm, significantly shorter compared to those planted in June, which ranged from 18.6 cm to 21.5 cm (Table 1).

2.1.2. Leaf Shape Index

A higher leaf shape index indicates longer and narrower leaves, while a lower inde[x] indicates shorter and wider leaves. The leaf shape index for tomatoes planted in April wa[s] significantly the highest, ranging from 1.14% to 1.29%. Tomatoes planted in May had [a] leaf shape index ranging from 0.96% to 1.19%, and those planted in June had a leaf shap[e] index ranging from 0.96% to 1.15%. In conclusion, tomatoes planted in June showed th[e] most vigorous growth, while those planted in April exhibited reproductive growth wit[h] less overall growth (Table 1).

2.1.3. Stem Diameter and Flowering Position

Tomatoes planted in April had a stem diameter ranging from 8.9 mm to 10.2 mm and [a] flowering position ranging from 14.4 cm to 21.3 cm (Table 2). Tomatoes planted in May ha[d] a stem diameter ranging from 9.5 mm to 11.9 mm and a flowering position ranging fro[m] 22.1 cm to 27.6 cm, indicating a weak vegetative growth type. Tomatoes planted in Jun[e] had a stem diameter ranging from 12.0 mm to 13.4 mm and a flowering position rangin[g] from 28.0 cm to 30.5 cm, indicating strong vegetative growth.

2.2. Yield Variation by Planting Time

The yield per unit area (m^2) of tomatoes was measured based on the planting tim[e]. Tomatoes planted in April yielded 12.57 kg/m^2; those planted in May yielded 10.8 kg/m^2 and those planted in June yielded 9.43 kg/m^2. April-planted tomatoes had the highes[t] yield, while those planted in June had the lowest. The weekly cumulative yield differenc[e] observed from 75 to 89 days after planting persisted until the end of the harvest perio[d] (Figure 1a). The cumulative solar radiation from the planting date to the last harves[t] date and the light use efficiency for fruit production (LUE_F: g·MJ^{-1}) were compare[d]. The cumulative solar radiation for April was 3157.3 MJ/m^2, with a LUE_F of 3.53 g·MJ^{-1}. For May, the cumulative solar radiation was 2911.7 MJ/m^2, with a LUE_F of 3.08 g·MJ^{-1}. For June, the cumulative solar radiation was 2673.6 MJ/m^2, with a LUE_F of 2.47 g·MJ^{-1}. Additionally, the LUE_F for April was 3.53 g·MJ^{-1}, which was 14.6% higher than May an[d] 42.9% higher than June (Figure 1b).

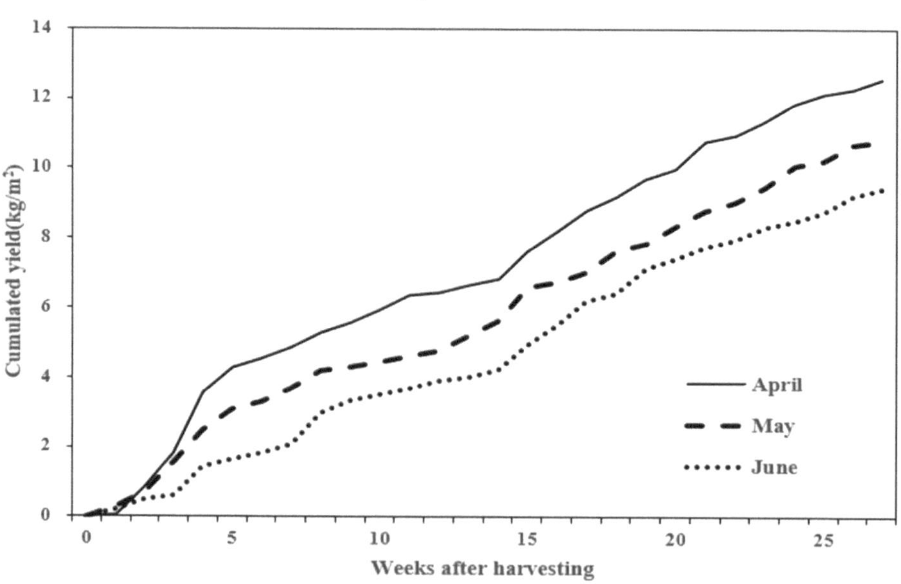

(a)

Figure 1. *Cont.*

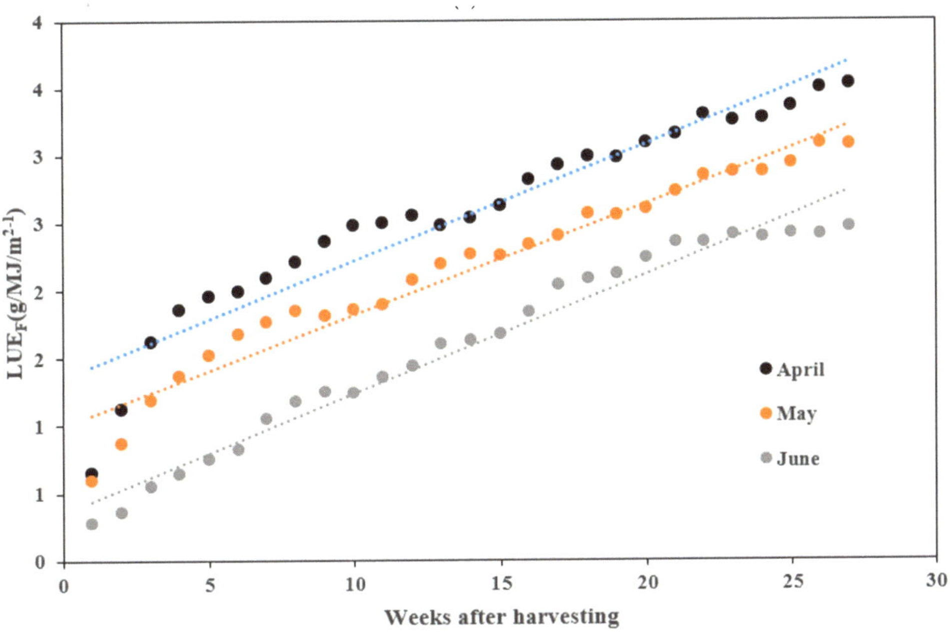

(**b**)

Figure 1. Comparison of harvest yield by planting period. (**a**) Cumulative harvest yield; (**b**) light use efficiency for fruit production (LUE_F).

2.3. Chlorophyll Fluorescence Response

Given that chlorophyll fluorescence is influenced by current weather conditions, the PI_{ABS}, ET_0/RC, and DI_0/RC results measured on the same day were compared (Table 3 and Figure 2).

PI_{ABS}: This index integrates the absorption capacity, electron transport efficiency, and electron trapping efficiency of PSII, reflecting overall photosynthetic activity [22]. The PI_{ABS} values for April were 3.677 during early growth and 3.468 during late growth (Table 3). For May, the values were 3.702 and 3.182, respectively, and for June, the values were 2.685 and 1.764, respectively. There was no significant difference between April and May, but June had significantly lower values, indicating reduced photosynthetic activity.

ET_0/RC: This index reflects the electron transport efficiency of PSII, with lower values indicating higher stress [23]. The ET_0/RC for April was 1.163 during early growth and 1.104 during late growth (Table 3). For May, the values were 1.163 and 1.124, respectively, and for June, the values were 0.999 and 0.858, respectively. There was no significant difference between April and May, but June had significantly lower values, indicating higher stress.

DI_0/RC: This index indicates the amount of absorbed light energy that is not used for photosynthesis but is lost as heat [24]. The early growth DI_0/RC values were 0.439 for April, 0.432 for May, and 0.444 for June, showing no significant differences (Table 3). However, during late growth, the values were 0.464 for April, 0.455 for May, and 0.637 for June, with June showing significantly higher values.

Table 3. Results of chlorophyll fluorescence measurement according to planting period.

	Planting Season	Days after Planting					
		Early Growth			Late Growth		
		96	124	152	209	237	265
PI_{ABS}	April	2.978 b	4.198 a	3.677 a	4.661 a	3.030 a	3.468 c
	May	5.274 a	3.702 b	2.423 b	3.277 b	3.182 a	4.135 b
	June	2.685 b	2.175 c	3.450 a	1.764 c	2.183 b	5.012 a
	LSD (5%) ^{y}	**	***	**	***	**	***
ET_0/RC	April	1.044 b	1.052 b	1.163 a	1.047 a	0.815 b	1.104 a
	May	1.140 a	1.163 a	0.803 c	0.790 c	1.124 a	1.069 a
	June	0.999 c	0.753 c	0.888 b	0.858 b	0.813 b	1.087 a
	LSD (5%) ^{y}	***	***	***	***	**	Ns
DI_0/RC	April	0.427 b	0.371 b	0.439 a	0.354 b	0.350 b	0.464 a
	May	0.358 c	0.432 a	0.428 a	0.308 b	0.455 a	0.375 b
	June	0.444 a	0.418 a	0.353 b	0.637 a	0.456 a	0.350 b
	LSD (5%) ^{y}	***	**	**	**	**	**
			Investigated: 1 August 2022			Investigated: 1 December 2022	

^{y} Mean separation within columns by Duncan's multiple range test at $p = 0.05$. **, and ***: significant at $p \leq 0.01$ and 0.001, respectively. The letters a, b, and c indicate the presence or absence of significant variability. The same letters shared between groups signify no significant differences, while different letters indicate significant variability ($p < 0.05$) among the groups.

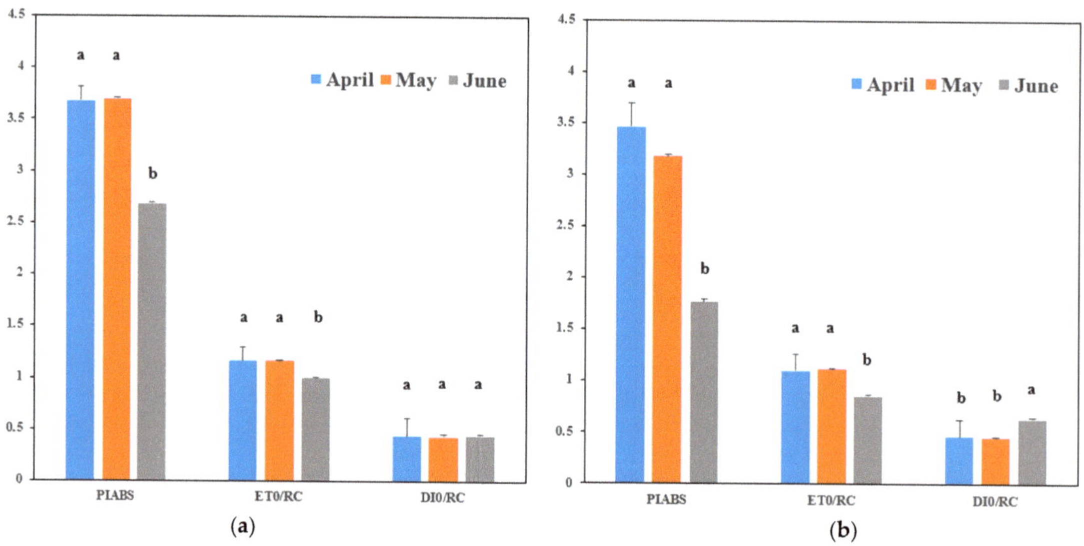

Figure 2. Chlorophyll fluorescence measurements according to tomato planting times. (**a**) Measurements during the early growth stage; (**b**): measurements during the late growth stage. The letters a and b indicate the presence or absence of significant variability. Bars with the same letters within each group signify no significant difference, while different letters indicate presence of significant variability ($p < 0.05$) among the groups.

3. Discussion

3.1. Growth Analysis Based on the Transplanting Time

Understanding the optimal planting time for tomato cultivation is crucial for maximizing growth, photosynthetic efficiency, and, consequently, fruit yield [25]. Environmental parameters such as relative humidity and CO_2 concentration were carefully controlled to ensure stable conditions conducive to accurate observation and analysis. The analysis focused on growth differences observed among tomatoes transplanted at different times. In tomatoes, stem diameter and flowering position are important physiological traits that reflect the plant's developmental stage and reproduction strategy and, subsequently, influence fruit setting and yield potential [26]. Significant variations in stem diameter and flowering position were evident across the planting times, indicating distinct growth patterns. Tomatoes planted in June exhibited the largest growth, characterized by thicker stems and higher positioning of flowers, indicative of strong vegetative growth. In contrast, April-planted tomatoes showed smaller stem diameters and lower flower positions, suggesting a reproductive growth type with less overall vigor. Flowering position is the distance from the epical meristem to the tip of the inflorescence. A shorter distance means that more flowers bloomed faster, which leads to more fruit, higher yields, and increased assimilation distribution to the fruit.

3.2. Yield Variation by Planting Time

Yield is an indispensable trait in tomato cultivation and crucial for assessing the overall productivity and economic viability. However, tomato yield can be influenced by planting time during specific seasons of cultivation [27]. The current study also assessed tomato yield across different planting times. April-planted tomatoes yielded the highest at 12.57 kg/m^2, followed by May-planted tomatoes at 10.8 kg/m^2, and June-planted tomatoes at 9.43 kg/m^2. This yield trend persisted throughout the harvesting period, indicating a consistent advantage for April-planted tomatoes despite variations in solar radiation and light use efficiency. The higher yield recorded in April-planted tomatoes can be attributed to their initiation of fruit setting before the onset of monsoon conditions, whereas June-planted tomatoes experienced lower initial fruit set due to adverse weather during their vegetative phase. It has been shown that tomato yield increases linearly with the increase in cumulative solar radiation [28,29], implying that tomato plants benefit from longer exposure to sunlight during their critical growth stages [30]. This extended exposure enhances photosynthesis, which is essential for robust vegetative growth, flower production, and, ultimately, fruit development [31]. Therefore, April-planted tomatoes, starting their fruit setting earlier in the summer season, not only avoid the detrimental effects of monsoon weather but also capitalize on a longer period of favorable solar radiation. This prolonged exposure to sunlight allows them to accumulate more energy and nutrients, translating into higher yields compared to tomatoes planted later in the season, such as in June. The summer climate in Korea, characterized by high temperatures, low light, and humid monsoons, reduces net photosynthesis and increases stress on crops. Tomatoes planted in April initiated their fruit setting before the onset of the monsoon, whereas those planted in June likely experienced low initial fruit set due to the onset of the monsoon during their vegetative growth phase, leading to reduced yield. Tomatoes planted in April yielded 1.77 kg/m^2 more than those planted in May and 3.14 kg/m^2 more than those planted in June. However, tomatoes planted in April had a stem diameter ranging from 8.9 mm to 10.2 mm lower than the diameters recorded for May and June. These results suggest that while April-planted tomatoes yield higher due to their early fruit setting and extended exposure to beneficial solar radiation, they may exhibit comparatively thinner stems. The thinner stem diameter observed in April-planted tomatoes could indicate potential trade-offs between structural integrity and reproductive output, influencing their overall resilience and support needs throughout their growth cycle.

3.3. Chlorophyll Fluorescence Response

A comprehensive understanding of tomato physiological responses can also be es tablished based on chlorophyll fluorescence measurements [22,32]. The absorbed light in plants is utilized in various ways. When light is absorbed, electrons are excited to a highe energy level within plant cells [33]. This elevated energy level must be reduced quickly to prevent cellular stress. Plants dissipate their absorbed light energy in three primary ways: through photosynthesis, emission as fluorescence by photosystem II (PSII), or a heat [34,35]. Increased stress reduces the efficiency of photosynthesis and fluorescence emission, resulting in more energy being dissipated as heat, which raises leaf tempera ture [35]. In this study, chlorophyll fluorescence measurements provided insights into the photosynthetic activity and stress responses of tomatoes under different planting times The photosynthetic performance, indicated by PI_{ABS} values, was highest for April-planted tomatoes and significantly lower for June-planted tomatoes during both the early and late growth stages. Similarly, indices reflecting electron transport efficiency (ET_0/RC) and energy dissipation as heat (DI_0/RC) confirmed higher stress levels in June-planted toma toes, particularly during the late growth stages. These results emphasize the sensitivity of tomato physiology to planting time, with April having been found as the optimal time for maximizing photosynthetic efficiency and minimizing stress-induced energy loss. This further indicates that more energy was dissipated as heat in June-planted tomatoes, reflect ing higher stress and less efficient photosynthesis. These results reveal the importance o planting time on the growth, yield, and physiological stress responses of tomatoes in the semi-closed greenhouse environment. Thus, planting tomatoes in April results in the mos favorable growth and yield outcomes, while planting in June leads to increased stress and reduced productivity. The findings in this study contribute valuable data to the field o agricultural science, providing a basis for informed decision-making by farmers and re searchers aiming to enhance crop productivity and sustainability in greenhouse cultivation systems. By elucidating the impact of planting time on growth, photosynthetic efficiency and yield, this study also emphasizes the importance of strategic planting practices tailored to local environmental conditions, thereby optimizing tomato production.

4. Materials and Methods

4.1. Experimental Site and Greenhouse Design

The experiment was conducted from April to August in 2022 using an advanced digital greenhouse established in 2021 by the National Institute of Agricultural Sciences in Iseo-myeon, Wanju-gun, Jeollabuk-do (32°39′40″ N, 51°40′49″ E) (Figure 3). It is a semi-closed Venlo-type greenhouse structure, and the covering material used is PO film The entire greenhouse has a height of 8.3 m, a side height of 7.0 m, a width of 40 m, and a length of 120 m, consisting of five sections (24 m × 40 m each). Each greenhouse can individually control temperature, humidity, solar radiation, CO_2, nutrient supply, heating, cooling, and actuators. For heating and cooling, the greenhouse uses six 840 kW air heat pumps (ACHH040LET2, LG, Republic of Korea), which store thermal and cooling energy in a 300-ton thermal storage tank. This energy is then distributed using tube rails and Fan Coil Units (FCUs). Additionally, the air conditioning room is equipped with internal and external air conditioning windows and FCUs, which selectively intake external air based on the internal and external temperatures. The irrigation method utilized solar radiation proportional control and was supplied at intervals of 80 J/m^2 to 120 J/m^2. The amount supplied per time was 100 cc/plant to 150 cc/plant.

(a)

(b) (c)

Figure 3. Overview and interior of the advanced digital greenhouse. (**a**) An overview of the advanced digital greenhouse; (**b**) the interior of the greenhouse; and the (**c**) air conditioning room.

4.2. Plant Material and Treatment Conditions

The tomatoes transplanted were grafted seedlings, using 'Dafnis' (Syngenta Korea, Seoul, Republic of Korea) as the scion and 'Spider' (Takii Korea, Seoul, Republic of Korea) as the rootstock. The coir substrate (Daeyoung GS, Republic of Korea) used measured 1 m in length, 0.2 m in width, and 0.1 m in height, with a chip-to-dust ratio of 7:3. The substrate was moistened to an EC concentration of 2.5 dS/m. The greenhouse was divided into three sections at 8 m intervals, and 6-week-old tomato plants were transplanted on 20 April, 19 May, and 21 June, with 360 plants in each section. The planting density was 1.13 plants per square meter, with a bed spacing of 1.6 m. The target drainage EC was set to 3.5–4.0 dS/m; the pH was set to 5.5–6.0; and the drainage rate was set to 35–40%. Based on changes in the drainage EC and pH, the nutrient solution EC was maintained at 2.3–2.5 dS/m and the pH at 5.5–6.0.

4.3. Environmental Conditions during the Experiment

During the experimental period, the monthly average external temperature ranged from −0.3 °C to 28.0 °C, and the monthly cumulative solar radiation ranged from 176.7 MJ/m^2 to 532.4 MJ/m^2. The internal temperature of the greenhouse was maintained between 17.1 °C and 23.9 °C. The relative humidity ranged from 65.9% to 85.1% and the CO_2 concentration was controlled between 367 ppm and 520 ppm (Figure 4).

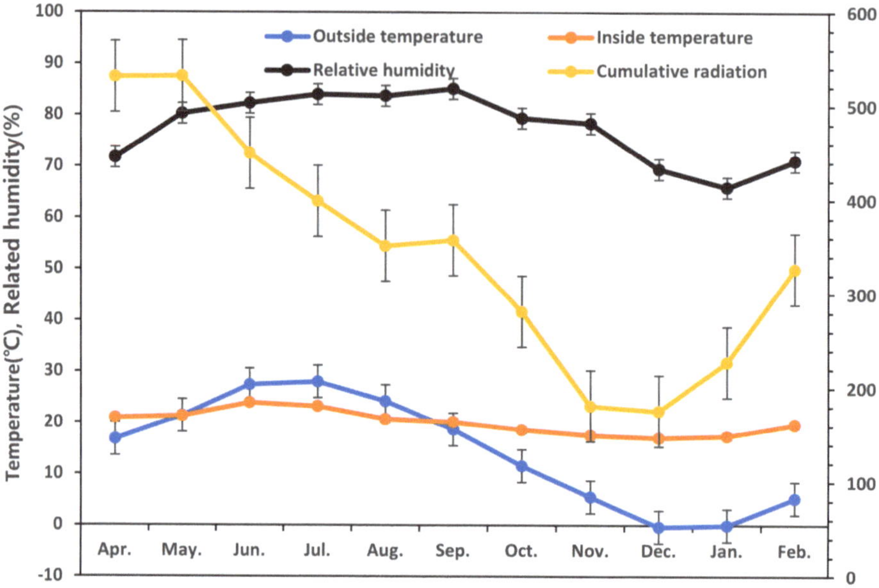

Figure 4. Average outside temperature (•), internal temperature (•), relative humidity (•), and total radiation (•) during the period of cultivation of tomatoes grown in greenhouses located in Jeonju.

4.4. Growth Assessment and Yield Measurement

To evaluate the growth and yield of tomatoes, 360 tomato plants were transplanted in each experimental stage, of which 36 plants were selected from the center for measurement. Growth comparisons were made during the early and late stages of growth over a three-week period. Early growth was evaluated from 75 to 89 days after transplanting, while late growth was assessed from 244 to 258 days after transplanting. The parameters measured included the stem diameter, flowering position, stem growth, leaf length, and leaf width. The leaf shape index was calculated by dividing the leaf length by the leaf width. Flowering position was determined by measuring the distance between the highest fully bloomed flower cluster and the growing point. Stem diameter was measured just below the flowering cluster. Stem growth was determined by measuring the distance from the mark on the support string, indicating the position of the growing point 7 days prior to the growing point. Leaf length and width were measured by assessing the horizontal and vertical dimensions of the lower leaves of the uppermost flowering cluster. Tomato fruit harvesting was conducted from 65 to 265 days after transplanting for each treatment, at intervals of 1 to 2 times per week. Fruits were harvested when they were more than 80% colored (red stage) and weighed using an electronic scale (KS-308, Dretec, Japan).

4.5. Chlorophyll Fluorescence Measurement

Light is closely related to crop yield, so chlorophyll fluorescence measurements were taken to compare the light use efficiency of crops planted at different times. Chlorophyll fluorescence was measured three times each during the early and late growth stages. The early growth measurements were taken at 95, 124, and 152 days after transplanting, while the late growth stage measurements were taken at 208, 237, and 264 days after transplanting. Measurements were conducted on the terminal leaves of the lower leaf of the highest flowering cluster for 36 plants in each treatment group. The leaves were dark-adapted for 30 min using a leaf clip, and chlorophyll fluorescence was measured using a Fluorpen FP-100 (Photon Systems Instruments, Brno, Czech Republic) to conduct the Fast chlorophyll a fluorescence induction (OJIP) test.

The OJIP test captures the following fluorescence signals:

- O phase: The minimum fluorescence measured at 50 μs when all photosystem II (PSII) reaction centers are open.
- J phase: Measured at 2 ms.
- I phase: Measured at 60 ms.
- P phase: The maximum fluorescence (F_M) measured when all PSII reaction centers are closed [24].

The data from the OJIP test were analyzed using FluorPen software (Photon Systems Instruments, 1.1.2.4 Version, Brno, Czech Republic). The indices derived from the OJIP test include F_0, F_M, F_V/F_M, PI_{ABS}, DI_0/RC, and ET_0/RC and are defined in Table 4.

- F_0: Indicates the energy emitted by chlorophyll molecules before being transferred to the PSII reaction center. An increase in F_0 signifies a greater number of inactive chlorophyll molecules, indicating reduced energy capture capacity [15].
- F_V/F_M: The most commonly used index in chlorophyll fluorescence analysis, representing the maximum photochemical efficiency of PSII [36]. It is calculated using the following formula: $F_V = F_M - F_0$.
- PI_{ABS}: An integrated index representing the absorption, electron transport efficiency, and electron trapping efficiency of PSII, reflecting overall photosynthetic activity [22,32].
- ET_0/RC: Reflects the electron transport efficiency of PSII, specifically the reduction of QA to QB. This index decreases under stress conditions [23].
- DI_0/RC: Indicates heat loss per reaction center (RC) at time zero. Energy not used for photosynthesis or emitted as fluorescence is dissipated as heat, which increases under stress [37].

Table 4. Description of OJIP parameters.

OJIP Parameters	Explanation
F_0	Minimal fluorescence yield of dark-adapted PS II
F_M	Maximal fluorescence yield of dark-adapted PS II
F_V	Maximal variable fluorescence ($F_V = F_M - F_0$)
PI_{ABS}	Performance index for energy conservation from photons absorbed by PSII antennas to the reduction of Q_B
F_V/F_M	Maximum quantum yield of primary PSII photochemistry
ET_0/RC	Electron transport (ET) flux from Q_A to Q_B per PS II reaction center (RC)
DI_0/RC	Heat dissipation (DI) at time zero per reaction center

4.6. Statistical Analysis

To compare the growth parameters, including the stem diameter, flowering position, stem growth, and leaf shape index, of tomatoes transplanted at different times in the semi-closed greenhouse, a statistical analysis was performed. The analysis was conducted using SPSS v.27 (IBM Corporation, Chicago, IL, USA). Duncan's Multiple Range Test (DMRT) was used to determine significant differences between treatments, with the confidence level set at 95%.

5. Conclusions

The results showed that the tomatoes planted in April had a stem diameter approximately 18.4% thinner than those planted in May and 34.7% thinner than those planted in June. Additionally, the flowering position of tomatoes planted in April was 66.7% higher than those planted in May and 44.0% higher than those planted in June. Moreover, stem growth in tomatoes planted in April was about 29.0% shorter compared to those planted in

May and 34.6% shorter compared to June. Reduced stem growth suggests a decrease in the frequency of stem lowering and guiding operations, potentially reducing labor costs. The leaf shape index for April plantings was about 11.9% higher compared to May and 14.6% higher compared to June, indicating a more balanced and weaker reproductive growth type. The cumulative solar radiation in April was 8.4% higher than in May and 18.1% higher than in June. The cumulative yield of tomatoes planted in April was 16.4% higher than those planted in May and 33.3% higher than those planted in June. Additionally, the light use efficiency for fruit production (LUE_F) in April was 14.6% higher than in May and 42.9% higher than in June. Therefore, April was identified as the period with the highest cumulative solar radiation and light use efficiency. Chlorophyll fluorescence measurements showed no significant differences between April and May. However, the PI_{ABS} values for June were 36.9% lower during early growth and 96.6% lower during late growth compared to April, indicating the lowest photosynthetic activity in June. ET_0/RC values for June were 16.4% lower during early growth and 28.7% lower during late growth compared to April. The DI_0/RC values for June showed no significant differences during early growth but were 37.3% higher during late growth compared to April and 40.0% higher compared to May, indicating higher stress levels and greater heat loss in June-planted tomatoes. Our analysis of growth, yield, and photosynthesis determined that April is the most suitable planting time for summer cropping in Korea. Planting in April can reduce labor costs and increase farm income due to higher yields. Future research should focus on the effects of supplementary lighting during the summer monsoon season in Korea, as this could further optimize growth conditions. Since this study was conducted on "Dafnis" and "Spider" grafted seedlings, results may vary for other varieties, highlighting the need for continued investigation across different tomato cultivars.

Author Contributions: Conceptualization, H.J.B. and I.C.; methodology, H.J.B. and I.C.; validation, H.J.B., Y.J. and I.C.; formal analysis, H.J.B. and Y.A.S.; investigation, H.J.B.; resources, H.J.B. and Y.J.; data curation, H.J.B., D.S.L. and B.K.; writing—original draft preparation, H.J.B.; writing—review and editing, S.-H.K., K.O. and Y.H.; visualization, H.J.B., S.-H.K. and S.P.; supervision, I.C. and J.H.B.; project administration, H.J.B. All authors have read and agreed to the published version of the manuscript.

Funding: This research was carried out with grant support from the Research Program for Agricultural Science and Technology Development (PJ01676001), National Institute of Agricultural Sciences, Rural Development Administration, Republic of Korea.

Data Availability Statement: All data are available within the paper.

Conflicts of Interest: The authors declare no conflicts of interest.

References

1. Kil, S.-H.; Park, H.-M.; Park, M.; Kim, Y.I.; Lee, E. Location Selection of Urban Rooftop Greenhouses in Seoul Based on AHP and GIS. *Land* **2023**, *12*, 2187. [CrossRef]
2. Kim, C.-G.; Lee, S.-M.; Jeong, H.-K.; Jang, J.-K.; Kim, Y.-H.; Lee, C.-K. *Impacts of Climate Change on Korean Agriculture and Its Counterstrategies*; Research Report R593; Korean Rural Economic Institute: Naju, Republic of Korea, 2010; pp. 1–306.
3. De Block, M.; Van Lijsebettens, M. Energy efficiency and energy homeostasis as genetic and epigenetic components of plant performance and crop productivity. *Curr. Opin. Plant Biol.* **2011**, *14*, 275–282. [CrossRef] [PubMed]
4. Cheong, D.-C.; Lee, J.-J.; Choi, C.-H.; Song, Y.-J.; Kim, H.-J.; Jeong, J.-S. Growth and cut-flower productivity of spray rose as affected by shading method during high temperature period. *Hortic. Sci. Technol.* **2015**, *33*, 227–232. [CrossRef]
5. Walker, D.; Sivik, M.; Prinsley, R.; Cheesbrough, J. Effect of temperature on the rate of photorespiration in several plants. *Plant Physiol.* **1983**, *73*, 542–549. [CrossRef] [PubMed]
6. In-hee, L. Change of rural development policy in South Korea after Korean war. *J. Reg. City Plan.* **2021**, *32*, 130–149. [CrossRef]
7. Won, J.; Babu, S.C.; Rana, A. *Building a Climate Change-Resilient Food System in Korea: The Case of Extension and Technology Dissemination Services*; International Food Policy Research Institute: Washington, DC, USA, 2019; Volume 1861.
8. Lee, D.; Kim, K. National investment framework for revitalizing the R&D collaborative ecosystem of sustainable smart agriculture. *Sustainability* **2022**, *14*, 6452. [CrossRef]
9. Sutthiwaree, P.; Lee, H.; Kwak, K.; Lee, S.; Kim, B. The Situation Development route of greenhouse in Thailand and Republic of Korea. *Thai Agric. Res. J.* **2020**, *38*, 90–108.

10. García-Vázquez, F.; Ponce-González, J.R.; Guerrero-Osuna, H.A.; Carrasco-Navarro, R.; Luque-Vega, L.F.; Mata-Romero, M.E.; Martínez-Blanco, M.D.R.; Castañeda-Miranda, C.L.; Díaz-Flórez, G. Prediction of Internal Temperature in Greenhouses Using the Supervised Learning Techniques: Linear and Support Vector Regressions. *Appl. Sci.* **2023**, *13*, 8531. [CrossRef]
11. Park, M.-L.; Jeong, J.-W. Review of Semi-closed Greenhouse Previous Studies and Required Technology Analysis for Improving Energy Efficiency. *KIEAE J.* **2022**, *22*, 21–27. [CrossRef]
12. Sapounas, A.; Katsoulas, N.; Slager, B.; Bezemer, R.; Lelieveld, C. Design, control, and performance aspects of semi-closed greenhouses. *Agronomy* **2020**, *10*, 1739. [CrossRef]
13. Weerakkody, P.; Amarathunga, S.; Warnakulasooriya, S.; Herath, S. Natural ventilation in insect screened single span greenhouses under warm weather. *Int. J. Biol. Chem. Sci.* **2009**, *3*, 1297–1309. [CrossRef]
14. De Zwart, H. Overall energy analysis of (semi) closed greenhouses. In Proceedings of the International Symposium on High Technology for Greenhouse System Management: Greensys, Naples, Italy, 4–6 October 2007; Volume 801, pp. 811–818.
15. Baker, N.R.; Rosenqvist, E. Applications of chlorophyll fluorescence can improve crop production strategies: An examination of future possibilities. *J. Exp. Bot.* **2004**, *55*, 1607–1621. [CrossRef] [PubMed]
16. Qian, T.; Dieleman, J.; Elings, A.; De Gelder, A.; Marcelis, L.; Van Kooten, O. Comparison of climate and production in closed, semi-closed and open greenhouses. In Proceedings of the International Symposium on High Technology for Greenhouse Systems: GreenSys, Quebec City, QC, Canada, 14–19 June 2009; Volume 893, pp. 807–814.
17. Le Quillec, S.; Albert, B.; Lesourd, D.; Loda, D.; Barette, R.; Brajeul, E. Benefits of a semi-closed greenhouse for tomato production in the West of France. In Proceedings of the International Symposium on New Technologies and Management for Greenhouses-GreenSys, Evora, Portugal, 19–23 June 2015; Volume 1170, pp. 883–888.
18. Lee, H.-K.; Lee, M.-H.; Park, G.-S.; Lee, E.-M.; Jeon, N.-B.; Seo, S.-D.; Cho, P.-H.; Kim, Y.-S.; Kim, S.-E.; Cho, S.-K. Effect of seedling type and early transplanting of summer grown seedling on the growth and yield of tomato. *Korean J. Org. Agric.* **2015**, *23*, 59–66. [CrossRef]
19. Kim, S.E.; Kim, Y.S. Optimum management of tomato side stems pruning in summer cultivation. *J. Bio. Environ. Control* **2014**, *23*, 167–173. [CrossRef]
20. Kim, E.J.; Park, K.S.; Goo, H.W.; Park, G.E.; Myung, D.J.; Jeon, Y.H.; Na, H. Effect of cooling in a semi-closed greenhouse at high temperature on the growth and photosynthesis characteristics in Paprika. *J. Bio. Environ. Control* **2021**, *30*, 335–341. [CrossRef]
21. Junker, A.; Muraya, M.M.; Weigelt-Fischer, K.; Arana-Ceballos, F.; Klukas, C.; Melchinger, A.E.; Meyer, R.C.; Riewe, D.; Altmann, T. Optimizing experimental procedures for quantitative evaluation of crop plant performance in high throughput phenotyping systems. *Front. Plant Sci.* **2015**, *5*, 770. [CrossRef] [PubMed]
22. Thach, L.B.; Shapcott, A.; Schmidt, S.; Critchley, C. The OJIP fast fluorescence rise characterizes Graptophyllum species and their stress responses. *Photosynth. Res.* **2007**, *94*, 423–436. [CrossRef] [PubMed]
23. Yoo, S.-Y.; Lee, Y.-H.; Park, S.-H.; Choi, K.-m.; Park, J.-Y.; Kim, A.-R.; Hwang, S.-M.; Lee, M.-J.; Ko, T.-S.; Kim, T.-W. Photochemical response analysis on drought stress for red pepper (*Capsiumannuum* L.). *Korean J. Soil Sci. Fertil.* **2013**, *46*, 659–664. [CrossRef]
24. Thwe, A.A.; Kasemsap, P. Quantification of OJIP fluorescence transient in tomato plants under acute ozone stress. *Agric. Nat. Resour.* **2014**, *48*, 665–675.
25. Shamshiri, R.R.; Jones, J.W.; Thorp, K.R.; Ahmad, D.; Man, H.C.; Taheri, S. Review of optimum temperature, humidity, and vapour pressure deficit for microclimate evaluation and control in greenhouse cultivation of tomato: A review. *Int. Agrophysics* **2018**, *32*, 287–302. [CrossRef]
26. Zuo, Z.; Tan, J.; Mao, H.; Zhang, X.; Qin, L.; Lv, T.; Zhuo, M. Modelling of tomato stem diameter growth rate based on physiological responses. *Pak. J. Bot.* **2017**, *49*, 1429–1434.
27. Ahammad, K.; Siddiky, M.; Ali, Z.; Ahmed, R. Effects of planting time on the growth and yield of tomato varieties in late season. *Progress. Agric.* **2013**, *20*, 73–78. [CrossRef]
28. Myung, D.J.; Bae, J.H.; Kang, J.G.; Lee, J.H. Relationship between radiation and yield of sweet pepper cultivars. *J. Bio. Environ. Control* **2012**, *21*, 243–246.
29. Joung, K.H.; Jin, H.J.; An, J.U.; Yoon, H.S.; Oh, S.S.; Lim, C.S.; Um, Y.C.; Kim, H.D.; Hong, K.P.; Park, S.M. Analysis of Growth Characteristics and Yield Pattern of 'Cupra' and 'Fiesta' Paprika for Yield Prediction. *J. Bio. Environ. Control* **2018**, *27*, 349–355. [CrossRef]
30. Izzo, L.G.; Mele, B.H.; Vitale, L.; Vitale, E.; Arena, C. The role of monochromatic red and blue light in tomato early photomorphogenesis and photosynthetic traits. *Environ. Exp. Bot.* **2020**, *179*, 104195. [CrossRef]
31. Paradiso, R.; Proietti, S. Light-quality manipulation to control plant growth and photomorphogenesis in greenhouse horticulture: The state of the art and the opportunities of modern LED systems. *J. Plant Growth Regul.* **2022**, *41*, 742–780. [CrossRef]
32. Strasser, R.J.; Srivastava, A.; Tsimilli-Michael, M. The fluorescence transient as a tool to characterize and screen photosynthetic samples. *Probing Photosynth. Mech. Regul. Adapt.* **2000**, *25*, 445–483.
33. Gao, H.; Wu, F. Physiological and transcriptomic analysis of tomato in response to sub-optimal temperature stress. *Plant Signal. Behav.* **2024**, *19*, 2332018. [CrossRef] [PubMed]
34. Fernández-Calleja, M.; Monteagudo, A.; Casas, A.M.; Boutin, C.; Pin, P.A.; Morales, F.; Igartua, E. Rapid on-site phenotyping via field fluorimeter detects differences in photosynthetic performance in a hybrid—Parent barley germplasm set. *Sensors* **2020**, *20*, 1486. [CrossRef]

35. Murchie, E.H.; Lawson, T. Chlorophyll fluorescence analysis: A guide to good practice and understanding some new application *J. Exp. Bot.* **2013**, *64*, 3983–3998. [PubMed]
36. Yoo, S. *Molecular Physiological and Photochemical Responses on Water Stress and Development of Irrigation Standard Based on Meteor logical Interpretation for Red Pepper (*Capsicum annuum *L.)*; The Graduate School of Biotechnology and Information Technolog Hankyong National University: Anseong, Republic of Korea, 2012; pp. 34–35.
37. Nussbaum, S.; Geissmann, M.; Eggenberg, P.; Strasser, R.J.; Fuhrer, J. Ozone sensitivity in herbaceous species as assessed by direc and modulated chlorophyll fluorescence techniques. *J. Plant Physiol.* **2001**, *158*, 757–766. [CrossRef]

plants

MDPI

Review

Low-Potassium Fruits and Vegetables: Research Progress and Prospects

Jiawei Cui †, Yongxue Zhang †, Hongmei Zhang, Haijun Jin, Lizhong He, Hong Wang, Panling Lu, Chen Miao, Jizhu Yu * and Xiaotao Ding *

Shanghai Key Laboratory of Protected Horticultural Technology, Horticulture Research Institute, Shanghai Academy of Agricultural Sciences, Shanghai 201403, China; cuijiawei@saas.sh.cn (J.C.); xuezylemon@foxmail.com (Y.Z.); zhanghongmei@saas.sh.cn (H.Z.); jinhaijun@saas.sh.cn (H.J.); hlznd02@163.com (L.H.); wanghong@saas.sh.cn (H.W.); lpl2245@163.com (P.L.); miaochen@saas.sh.cn (C.M.)

* Correspondence: yy2@saas.sh.cn (J.Y.); dingxiaotao@saas.sh.cn (X.D.)

† These authors contributed equally to this work.

Abstract: With the increasing number of patients with chronic kidney disease (CKD) and the improved recognition of nutritional therapy, research on low-potassium (LK) fruits and vegetables for CKD patients has gained global attention. Despite its already commercial availability primarily in Japan, public awareness remains limited, and cultivation methods lack a comprehensive strategy. This review offers an extensive examination of the developmental significance, current cultivation techniques, and existing limitations of functional LK fruits and vegetables with the objective of providing guidance and inspiration for their exploitation. Additionally, this review investigates various factors influencing K content, including varieties, temperature, light, exogenous substances, harvest time, and harvest parts, with a focus on optimizing production methods to enhance potassium utilization efficiency (KUE) and decrease the K content in plants. Finally, the review outlines the shortcomings and prospects of research on LK fruits and vegetables, emphasizing the importance of interdisciplinary research (in agriculture technology, medicine, and business) for patients with CKD and the future development of this field.

Keywords: chronic kidney disease (CKD); hyperkalemia; soilless culture; potassium utilization efficiency (KUE); tolerance to low-potassium (LK) stress

Citation: Cui, J.; Zhang, Y.; Zhang, H.; Jin, H.; He, L.; Wang, H.; Lu, P.; Miao, C.; Yu, J.; Ding, X. Low-Potassium Fruits and Vegetables: Research Progress and Prospects. *Plants* **2024**, *13*, 1893. https://doi.org/10.3390/plants13141893

Academic Editor: Ferenc Fodor

Received: 17 May 2024
Revised: 28 June 2024
Accepted: 7 July 2024
Published: 9 July 2024

1. Introduction

Potassium ions (K^+) are the most abundant cation in plants (non-halophytes) [1]. They are involved in a variety of plant functions, including osmoregulation and cell extension, stomatal movement, activation of enzymes, protein synthesis, photosynthesis, stress responses, phloem loading, and phloem transport and uptake [2,3]. Adequate K supply maintains the normal root morphology [4], enhances water and nutrient uptake [4,5], improves the photosynthetic efficiency [6], promotes the growth of plants [7], and meliorates the quality of fruits [8]. Although agriculture generally maintains high yield and high quality in fruits and vegetables by increasing the potassium content in plants [7,9], low-potassium (LK) fruits and vegetables for chronic kidney disease (CKD) patients have recently become a research hotspot [10,11].

CKD has become a significant global health concern, affecting an estimated 8–13% of the global population [12]. CKD is prevalent among the elderly, and with the increasing aging of society in certain countries, its prevalence is rising [13]. In CKD patients with impaired renal function, there is a decreased capacity to excrete K, leading to an increased risk of hyperkalemia [14]. Hyperkalemia causes symptoms such as a slow heart rate, numbness and soreness of limbs, and, in the worst cases, may lead to flaccid hemiplegia or cardiac arrest [15]. Consequently, CKD patients are advised to restrict high-K foods, with a recommended daily K intake below 3 g [16]. Fresh fruits and vegetables typically have a high K content. However, restricting their consumption can lead to deficiencies i essential vitamins and minerals [17], as well as nutrients rich in fiber and alkali, whic can result in constipation and metabolic acidosis, both of which are significant risk factor for hyperkalemia [18]. Incorporating LK fruits and vegetables into CKD patient diet offers a solution to balancing nutritional needs while limiting K intake [19]. Therefor investigations and applications of functional LK fruits and vegetables hold significan promise for individuals with CKD. Currently, there is no universally acknowledged I value for classifying fruits and vegetables as LK. In Japan, fruits and vegetables with K content below 100 mg per 100 g of fresh weight (FW) are considered LK fruits an vegetables [20].

Investigations of LK fruits and vegetables for patients with CKD originated in Japan [21 Various studies have reported research on LK fruits and vegetables, encompassing spinach [10 lettuce [22], strawberry [23], melon [24], onion [11] and tomato [25]. The Aizu-Wakamats Akisai Vegetable Plant Factory of Fujitsu has achieved a notable production of 2500 head of LK lettuce daily, establishing itself as Japan's largest plant factory for cultivating LI vegetables [20]. Studies on LK fruits and vegetables conducted in other countries such a China [26], South Korea [27], and Italy [19] have also been documented.

In this review, we present various existing methods for producing LK fruits an vegetables, outline the factors affecting the K content in plants, and suggest avenues for th further optimization of production methods.

2. Cultivation Methods of LK Fruits and Vegetables

Because the mineral elements in the nutrient solution of soilless cultivation are highl controllable, the production of LK fruits and vegetables primarily relies on soilless cult vation. This can be achieved by reducing the supply of K to the nutrient solution [26,28 Production methods are categorized into two types based on nutrient solution managemen LK supply via electrical conductance management (LKEC) and LK supply via nutrien quantitative management (LKQM) [28].

2.1. LKEC Method

The LKEC method refers to the reduction in the K supply while maintaining the EC of the nutrient solution during the cultivation of fruits and vegetables [28]. Conductivit managements are commonly used for fertilizer and water control in soilless cultivatior This involves establishing a target EC based on plant fertilizer demand. Fertilizers ar added when the nutrient solution falls below the target EC value, whereas water can b added when it exceeds this value. The LKEC method operates on the principle of managin the nutrient solution EC.

The LKEC method encompasses three adjustment approaches to the nutrient solutio formula. The first method involves reducing the application of K nitrate (KNO_3) (th KNO_3 reduction method). The second method maintains the levels of other minera elements while only decreasing the K element (single-K element reduction method). Fo example, KNO_3 in formula fertilizers is substituted with ammonium nitrate (NH_4NO_3 The third method maintains the nitrate ion (NO_3^-) content while substituting the K ion (K^+) with sodium ions (Na^+) or calcium ions (Ca^{2+}) (cation substitution method). Fo example, KNO_3 is replaced with sodium nitrate ($NaNO_3$) or calcium nitrate ($Ca(NO_3)_2$) ir formula fertilizers.

The first LKEC option is the KNO_3 reduction method. K in the soilless cultivatior nutrient solutions is commonly supplied as KNO_3. Therefore, reducing the applicatior of KNO_3 could decrease the supply of K. Asaduzzaman [24] observed that melon plant in perlite culture, receiving only 50% of the required KNO_3 during the 3rd and 4th week after transplantation, produced fruits with 53% less K content than the control group The resulting LK melon was well received by patients with CKD owing to its appealin taste and the absence of a tingling sensation. Fuad Mondal [23] attempted to produce LK strawberry fruits by reducing the KNO_3 fertilizer in the nutrient solution from anthesis tc harvest. The 'Toyonoka' variety exhibited a 64% reduction in K content when cultivated in a nutrient solution containing KNO_3 at 1/16 of the standard concentration.

The second LKEC option is the single-K element reduction method. Substituting NH_4NO_3 for KNO_3 can reduce the K element while maintaining the nitrogen levels. This approach decreases the K content in vegetables. Liu [26] utilized NH_4NO_3 to replace a portion of KNO_3 in the formula. As the K concentration in the nutrient solution decreased, the K content in the mustard, lettuce, and spinach exhibited a significant decline. The mustard, lettuce, and spinach treated with 1/8 K exhibited K contents 79.93%, 66.37%, and 60.33% lower than that of the normal K concentration, respectively. Moreover, the 1/8 K treatment presented higher levels of soluble solids and soluble sugar than the control group, while nitrogen, phosphorus, Ca and magnesium levels remained similar to the control group [26].

Last, there is the cation substitution method. Studies have indicated that Na^+ and Ca^{2+} can potentially substitute for K^+ in certain nonspecific physiological functions, such as osmoregulation [29,30]. This method of substituting K^+ with Na^+ or Ca^{2+} can mitigate the stress induced by LK levels in plants and concurrently reduce the K content. At present, the method of using $Ca(NO_3)_2$ or $NaNO_3$ to replace part of the KNO_3 in the nutrient solution has successfully produced kale [27], lettuce [31] and microgreens [19] and other vegetables with low potassium content, and these vegetable yields are no different from the control. Son et al. [27] demonstrated this by substituting KNO_3 with $Ca(NO_3)_2$ 3 weeks after the kale planting, maintaining the same EC value. After 3 weeks of K deprivation treatment, the K content of kale demonstrated a 70% reduction in K content without any adverse effects on yield or visual quality. Additionally, an increase in the health-promoting component, glucosinolates, was observed. Renna et al. [19] successfully reduced the K content in seedling vegetables by using $NaNO_3$ to replace either all or 75% of KNO_3 in the nutrient solution. This resulted in seedling vegetables containing only 103–129 mg of K per 100 g of FW compared to the normal K supply of 225–250 mg K per 100 g of FW. The K content of the three Astericae seedlings ('Molfetta', 'Italico a costa rossa', and 'Bionda da taglio') decreased by approximately 50%, and the FW slightly decreased (17.5%). There were no significant differences in plant height, dry matter, or commercial value. Furthermore, there were no negative effects on the nutritional quality (total lipid, protein, total carbohydrate, fiber, ashes, and antioxidant activity).

Although all three methods can yield fruits and vegetables with LK content, the determination of the method should be based on the specific characteristics of the fruits and vegetables species. When employing the cation substitution method, consideration should be taken to the tolerance of fruits and vegetables to Na and Ca stress. Liu [26] compared the effects of the single-K element reduction method and the cation substitution method on the growth, mineral elements, and quality of three leafy vegetables (mustard, lettuce, and spinach). The results indicated that when the K supply level in the nutrient solution was below 1.19 mmol·L^{-1} for both methods, the K content in mustard and lettuce decreased by over 50%. The cation substitution method was more suitable for the LK production of mustard and lettuce, as it did not adversely affect yield. Conversely, the single-K element reduction method significantly reduced the yield of mustard and lettuce. Spinach could be sensitive to Na, and high Na levels inhibited the growth. Of both methods, the single K reduction method has little effect on the spinach yield, making it more suitable for spinach cultivation.

The timing of LK treatment could be crucial for implementing LKEC. Zhang et al. [22] demonstrated that initiating LK nutrient solution from the time of colonization led to reduced lettuce yield, rendering it ineffective for the LK lettuce production. However, Ogawa [31] has suggested that withholding K application during the latter half of the growth period for three types of leafy vegetables presented no difference in fresh and dry weights compared to the control group. This suggests that providing K normally during the early growth stage and then removing it during the late growth stage is preferred for LK production in fruits and vegetables [31]. K can be removed from the nutrient solution 1 or 2 weeks before harvesting for leafy vegetables [31], after flowering for solanaceous and cucurbitaceous plants [24], and at the bulb expansion stage for bulb vegetables [11].

2.2. LKQM Method

Researchers have employed the LKQM method to distribute the K requirements of plants across different stages of their growth [28]. A novel approach to nutrient management in hydroponics has emerged, in which fertilizer is regularly supplied to the cultivation system (quantitative management) regardless of the EC of the solution [20,32]. Despite fluctuations in nutrient levels with this method, researchers have suggested that the roots can efficiently absorb ions in large solution volumes without impeding plant growth [33]. This concept has been applied to the K supply for cultivating LK fruits and vegetables. Some researchers have successfully produced LK vegetables by using the LKQM method. In a LK cherry tomato cultivation experiment [25], the total quantity of K supply per plant during the cultivation was 7.2 g (1 K, set as control), 3.6 g (1/2 K), 1.8 g (1/4 K), 0.9 g (1/8 K) and 0.6 g (1/12 K), respectively, which means five quantitative treatments of K fertilizer were set in the experiment. Researchers allocated 25%, 50%, and 25% of the total designated K across three stages (before the first flower anthesis, after the first flower anthesis, and after the green maturity of the first truss) in respective treatments with the K amount evenly distributed for weekly application. When the total designed K was set at 1/4 K and 1/8 K, the average K content of fruits decreased from 151.8 mg·100 g^{-1} FW in the control to 107.6 mg·100 g^{-1} FW and 76.4 mg·100 g^{-1} FW, respectively. However, the fruit yield, sugar content, and acid content remained largely unchanged compared with the control. Okada et al. [11] investigated the amount of K absorbed at the onion bulb expansion stage under a sufficient amount of fertilizer. Then, they evenly distributed 1.5 times the amount of mineral elements absorbed during onion bulb expansion over the 15 weeks of the expansion stage, resulting in LK onions with a K content of 88.3 mg·100 g^{-1} FW, which was 41.1% lower than the value (150 mg·100 g^{-1} FW) reported in the Standard Tables of Food Composition in Japan (2015). Nevertheless, this had little impact on bulb yield.

The LKQM method appears to offer more advantages than the LKEC method. Xu et al. [28] conducted a comparison of their effects on lettuce growth and K content. They observed no significant differences in K content and plant growth between the LKEC and LKQM treatments. However, lettuce treated with LKQM demonstrated lower Na content than that treated with LKEC, and the overall fertilizer usage was reduced.

3. Optimization of the Cultivation Methods of LK Fruits and Vegetables

Currently, LK fruits and vegetables cultivation primarily focuses on controlling the K levels in nutrient solutions. However, the K content of fruits and vegetables is also affected by various factors, such as varieties [22], growing conditions [34], and exogenous substances [17]. Therefore, optimizing LK fruits and vegetables cultivation methods should consider these factors along with K regulation.

3.1. Utilizing High K Utilization Efficiency (KUE) Varieties

Varieties possessing high KUE genotypes offer significant advantages for LK fruits and vegetables cultivation. KUE refers to the dry matter production per unit of accumulated K, which is inversely related to the K content [35,36]. High KUE varieties typically exhibit a lower relative K content and greater tolerance to LK stress [37]. For example, among the lettuce varieties, crisphead demonstrated the lowest K content, 42.91% less than Romaine lettuce, because of its higher KUE [38]. Additionally, Yang [39] reported that K-efficient rice varieties (JNZ) yielded 1.59 times more grain than K-inefficient rice varieties (KQ47) under LK soil conditions.

Significant intraspecific variations in KUE have been documented across crops. Wu [40] examined K uptake and KUE among 56 barley varieties, revealing that the K dry matter production efficiency, K dry matter production index, and dry matter weight of the K-efficient genotype were 1.4–2.3, 2.1–3.9, and 1.7–2.1 times higher, respectively, than those of the K-inefficient genotype. The genotypic differences in KUE were primarily attributed to (1) the variances in K partitioning and redistribution at the cellular and whole plant levels, (2) the substitution ability of K by other ions [41], and (3) the proportion of resources allocated to economic products [36]. Table 1 illustrates the cultivars with high K efficiency and their phenotypes for certain crops. The K-efficient genotypes typically exhibit developed and dense roots [36,42–47], enhanced K uptake ability [40,44,45,48], efficient translocation and distribution of both K and carbohydrates [36,39,46], effective K recycling and reuse [37], greater Na substitution capacity [41], higher photosynthetic pigments [47], and a relative net photosynthetic rate under K deficiency [39]. Moreover, the K-efficient varieties demonstrate a higher grain-filling rate at maturity, increased grain weight per spike, and an elevated harvest index than K-inefficient varieties in rice [39] and wheat [49].

Table 1. K-efficient genotype cultivar and K-inefficient genotype cultivar of some plant varieties, and characteristics of K-efficient genotype cultivar compared with K-inefficient genotype cultivar.

Plant Variety	K-Efficient Genotype	K-Inefficient Genotype	Phenotype of K-Efficient Genotype Cultivar	References
Barley	Sandrime	AC Westech	Higher K uptake, K dry matter production index, K dry matter production, efficiency, and dry matter weight	[40]
Cotton	103	122	Developed root systems, higher LK tolerance, better nutrition uptake capability, and stronger transport organs	[44,45]
Pear	*Pyrus ussuriensis*	*Pyrus betulifolia*	Higher LK tolerance, more efficiently recycles and reuses K	[37]
Potato	Huayu 5, Zhengshu 5	08CA0710, 09307-830, B20-7, Liangshu 2	Lower leaf and stem K content	[50]
Proso millet	Var 87, Var 189	Var 116	Developed and dense roots, higher photosynthetic pigments, and higher LK tolerance	[47]
Rice	HA-88, EJF, JNZ	KQ47, 81-280, TLHZ	Greater efficient translocation and distribution of both K and carbohydrate; higher relative net photosynthetic rate under LK supply; greater relative tillering rate during the tillering stage and a greater relative grain-filling rate during the late grain-filling stage	[39]
Sweet potato	Nan 88, Xushu 18, Shang 52-7, Zhe 6025	Zi 892, Meiguohong, Zi 1	Higher root weight and root: top ratio and harvest index (HI), and lower K concentration and accumulation	[43]
Sweet potato	Xu28, Wan5	Ji 22	Higher relative root weight per plant, lower K concentrations in the roots or whole plants at maturity, and better K translocation in the shoots and roots	[36,46]
Tea plant	1511	1601	Developed root systems and higher LK tolerance	[42]
Tobacco	Qinyan 96, Yuyan 6, Yunyan 87, Cuibi 1, Zhongyan 100	Eyan 1, RG17, Honghua Dajinyuan, G28, K326	Lower leaf and stem K content, root vigor, and K^+ influxes	[51]
Tomato	576, 571	349, 203, 546	Higher Na substitution capacity	[41]

Table 1. *Cont.*

Plant Variety	K-Efficient Genotype	K-Inefficient Genotype	Phenotype of K-Efficient Genotype Cultivar	References
Watermelon	ZXG0516, ZXG1553, ZXG0620, YS	WFL, 8424, NBT, HJX1	Higher relative shoot dry weight (ratio between shoot growth at limited K and that at adequate K), higher K uptake ability, and lower K concentration	[48]
Wheat	Yunmei 5	94-18	Higher grain weight per spike and harvest index, lower stem K concentration at maturity	[49]

In LK fruits and vegetables cultivation, the standard practice involves providing a normal K supply in the early growth stages followed by K deficiency later on. Hence selecting varieties capable of efficiently reutilizing the absorbed K is paramount. However there is a dearth of research on this subject, making it a key area for future investigation of LK fruits and vegetables cultivation.

3.2. Optimize the Cultivation Temperatures

The temperature can significantly affect the plant growth and development, influencing the mineral absorption, distribution, transformation, and utilization efficiency [52,53]. Generally, the K uptake responds quadratically to temperature variations with the optimal temperatures for nutrient uptake and growth aligning closely [54–57]. Several studies have reported a positive correlation between K content and uptake. Increasing root zone temperature under lower air temperatures can enhance K uptake and elevate the shoot and root K content in crops such as tomatoes [58] and grapevines [59]. However, excessively high root-zone temperatures may hinder root development and reduce mineral absorption and transportation efficiency [60]. This can lead to decreased K content in plant leaves, as observed in various crops, such as salad rocket [34], cool-season grass [61], *Lactuca sativa* L. cv. *Panama* [62], and cucumber [63]. Nonetheless, the increase in biomass due to the appropriate temperatures may dilute the K in tissues, introducing uncertainty in the K content–temperature relationship. Studies have shown that the total K uptake in star fruit increases with increasing root temperature from 5 to 25 °C, with no significant difference in leaf K content [43], which is a finding corroborated by Boatwright [64]. Moreover some studies have suggested that high temperatures exert a more pronounced negative impact on the KUE than on absorption efficiency, resulting in higher K content at elevated temperatures than under optimal conditions [54,56,65].

Although harsh ambient temperatures may decrease plant K content, the challenge of slowing growth rates and diminished marketability for commercial production cannot be ignored. The author suggests exploring alternative regulatory methods to reduce K content instead of inducing temperature stress.

3.3. Optimize Light Conditions

Light and K play pivotal roles in various biological processes including stomatal movement and crop quality. As key elements of light signaling, factors such as light intensity, quality, and photoperiod also govern the K uptake and utilization in plants [66]. Recently, researchers have investigated the influence of different light conditions on K absorption and concentration in plants (Table 2).

Previous studies have indicated that plant K uptake initially increases and then decreases with increasing light intensity with higher uptake occurring under moderate light conditions [54,67]. However, increased light intensity enhances the KUE, resulting in the decreased K content. This trend has been observed across various plant species, including basil [68], brassica microgreens [69], lettuce [54,70], herbaceous plants [71], *Tulbaghia violacea* L. [72], and *Erythrophleum fordii* Oliv. [73].

In addition to the light intensity, the K absorption and concentration are also influenced by the spectrum, although this effect may vary across different species or environmental conditions [74]. The multi-band composite spectra have been shown to mitigate damage caused by monochromatic light in *Coriandrum sativum* L. [75] and cucumber [76], resulting in a higher KUE and a lower K content compared to monochromatic light conditions. However, red light conditions led to the lowest K content in Chinese chive [77] and garlic seedlings [78]. In contrast, the K content of lettuce and Michaelmas daisy remained constant regardless of the presence of monochromatic or compound light in other experiments [79–82]. The impact of the red/blue (R:B) ratio on mineral absorption and accumulation has been extensively studied, as red and blue LEDs are commonly used for indoor plant cultivation [83]. The sweet basil presented an initial increase and subsequent decrease in K uptake and content as the R:B ratio varied from 0.5 to 4 with a peak uptake observed at 3 and the lowest at 0.5 [83]. Although Chen [84] also observed a similar pattern in lettuce, the correlation between the K content and the R:B ratio yielded inconsistent results across experiments. Wu [85] noted a decreasing trend in K content in lettuce as R:B ratios ranged from 100:0 to 60:40, whereas Liu [86] observed an initial increase followed by a decrease in K content as R:B ratios ranged from 3:1 to 1:3 with the highest content at 1:1 and the lowest at 3:1. However, studies by Chen [84] and Tian [87] did not find a correlation between the K content and changes in the R:B ratio. Under the identical R:B ratios, Kopsell [88] indicated that different wavelengths of blue light impacted mineral nutrients in kale (*Brassica oleracea* var. *acephala*) microgreens with lower blue LED wavelengths correlating with decreased K content. Green light has been associated with a reduced K content in lettuce [85] and flowering Chinese cabbage [89]. Plants grown under fluorescent lights (FLs) or high-pressure sodium lights (HPSs) with more green light bands exhibited lower K content than those grown under LED lights containing only red and blue light, supporting the hypothesis that green light reduces K content in plants [83,90,91]. Far-red (Fr) light also influences K absorption and utilization in certain plants. An increased proportion of far-red light promotes K absorption in lettuce [91] and the accumulation of K content per unit in tomato seedlings [92] while decreasing the K content in Chinese kale [93] and spinach [94]. The supplementation with trace amounts of UV-A has been shown to enhance basil growth and improve KUE, consequently reducing K content [95]. Although the UV-B radiation can reduce K content in some plants, its application is not suitable for LK vegetable production because of its predominantly injurious effects on plant growth [96].

Extending the light duration appropriately decreases the plant K content in some crops. Prolonged light exposure enhances K^+ absorption by roots, its transport to the shoot, and the overall K accumulation per plant [97–100]. However, the relationship between K application efficiency and photoperiod varies among plants. Increasing the duration of light exposure generally enhances plant biomass accumulation while reducing K content. For example, Li and Liu [98] observed a reduction in cucumber shoot K content with three hours of nighttime light supplementation. Continuous lighting before harvest significantly decreases lettuce K content [86,99]. Han [101] reported that extending the photoperiod from 12 to 16 h increased the K content in leaves but decreased it in fruits. Furthermore, in experiments investigating light–dark rhythm effects on lettuce mineral content, researchers found lower K content with lower light–dark alternation frequencies [102].

In conclusion, utilizing specific light conditions to promote plant growth is effective in reducing K content, serving a dilution function [103]. This strategy is particularly advantageous for indoor cultivation, where artificial lighting is essential. Before the implementation, it is imperative to conduct experiments or refer to similar results from the literature to optimize light conditions for various plants to reduce K content.

Table 2. Differential light conditions affecting K uptake and concentration in various plant varieties

Light Condition		Variation Trend	Plant Variety	K Uptake	K Concentration	References
Light intensity		Went from 7 to 15 (DLI, mol·m^{-2}·d^{-1})	Basil		↓	[68]
		Went from 105 to 315 (PPFD, μmol·m^{-2}·s^{-1})	Brassica microgreens		↓	[69]
		Went from 120 to 240 (PPFD, μmol·m^{-2}·s^{-1})	Cucumber		↑	[104]
		Went from 100 to 600 (PPFD, μmol·m^{-2}·s^{-1})	Lettuce	First↑ then ↓, highest on 500	↓	[54]
		Went from 300 to 450 (PPFD, μmol·m^{-2}·s^{-1})	Lettuce		—	[87]
		Went from 150 to 450 (PPFD, μmol·m^{-2}·s^{-1})	Lettuce		↓	[70]
		Went from 65 to 446 (PPFD, μmol·m^{-2}·s^{-1})	New Guinea impatiens	Irregularity		[67]
		90% to 0% shading	*Dactylis glomerata*, *Festuca ovina*, *Trifolium subterraneum*, *Medicago lupulina*		↓	[71]
		15% to 100% full light	*Erythrophleum fordii* Oliv.		↓	[73]
		33% to 100% full sunlight	Tomato	↓		[105]
		40% to 0% shading	*Tulbaghia violacea* L.		↓	[72]
Light quality	Monochromatic light vs. compound light	R, B, G, 7R1B, 13R2B1Fr	*Coriandrum sativum* L.		The sequence from high to low: R, B, G, 13R2B1Fr, 7R1B	[75]
		R, B, 8R1B, W FL	Cucumber		The sequence from high to low: R, B, 8R1B, W	[76]
		R, B, 3R1B, 7R1B, W FL	Chinese chive		The sequence from high to low: B, 3R1B, 7R1B, W, R	[77]
		R, B, 3R1B, W FL	Garlic seedling		The sequence from high to low: B, 3R1B, W, R	[78]
		R, B, RB, UV-A, W FL	Lettuce (*Lactuca sativa* L. 'Lollo Rosa')		—	[81]
		R, B, RB, UV-A, W FL	Lettuce (*Lactuca sativa* L. 'Chung Chi Ma')		—	[82]
		R, B, 8R1B, W FL	Lettuce		—	[79]
		R, B, G, RB, YR, W FL	Michaelmas daisy		—	[80]

Table 2. *Cont.*

Light Condition		Variation Trend	Plant Variety	K Uptake	K Concentration	References
Light quality	R/B light ratio	R/B light ratio from 100:0 decreased to 0:100	Lettuce	First ↑ then ↓, highest on 70:30, lowest on 100:0	—	[84]
		R/B light ratio from 100:0 decreased to 60:40	Lettuce		↓	[85]
		R/B light ratio from 3:1 decreased to 1:3	Lettuce		First ↑ then ↓, highest on 1:1, lowest on 3:1	[86]
		R/B light ratio from 8:1 decreased to 1:1	Lettuce		—	[87]
		R/B light ratio from 4:1 decreased to 1:2	Sweet basil	First ↑ then ↓, highest on 3:1, lowest on 1:2	First ↑ then ↓, highest on 3:1, lowest on 1:2	[83]
		W LED, RB400, RB420, RB450, RB470	*Brassica oleracea* var. *acephala*		The sequence from high to low is: W, RB470, RB450, RB420, RB400	[88]
	G light	G light ratio rose	Lettuce		↓	[85]
		G/R light ratio rose	Flowering Chinese cabbage		↓	[89]
		W FL vs. RB	Broccoli		↓	[90]
		W FL vs. RB	Sweet basil		↓	[83]
		W HPS vs. RB	Lettuce		↓	[91]
	Far-red Light	Supplement far-red light	Chinese Kale		↓	[93]
		Supplement B, R, Fr LED to W FL	Lettuce		—	[106]
		Far-red/red light ratio rose	Lettuce	↑		[91]
		Supplement Fr to W LED light	Spinach		↓	[94]
		Supplement Fr	Tomato		↑	[92]
		Supplement Fr to R LED light	Tomato	—	—	[107]
	UV	Supplement UV-A	Basil		↓	[95]
		Supplement UV-B	Soybean seedlings		Root ↓, steam ↓, leaf ↑	[108]
		Supplement UV-B	Spring wheat		↑	[109]
		Supplement UV-B	Soybean		↓	[96]
		Supplement UV-B	Mono-maple seedlings		Root ↑, steam ↓, leaf ↑	[110]
		Supplement UV-B	Mung bean seedlings		Root ↑, steam ↓, leaf ↑	[111]

Table 2. *Cont.*

Light Condition	Variation Trend	Plant Variety	K Uptake	K Concentration	References
Photoperiod	Supplementary illumination duration from 0 h increased to 12 h	Cucumber	↑	Irregularity, lowest on 3 h	[98]
	12 h extends to 24 h	Lettuce	—	↓	[86]
	12 h extends to 16 h	Lettuce	↑		[100]
	Continuous lighting before harvest	Lettuce	↑	↓	[99]
	8/4 h (T12), 16/8 h (CK), 24/12 h (T36), 32/16 h (T48)	Lettuce	—	The sequence from high to low is: T12, T36, CK, T48	[102]
	Lengthen	Sweet Pepper	↑		[97]
	12 h extends to 16 h	Tomato		Leaf ↑, Fruit ↓	[101]

DLI: daily light integral; PPFD: photosynthetic photon flux density; LED: light-emitting diode; HPS: high-pressure sodium lamp; R, B, G, Y, Fr, UV-A, and UV-B represent red light LED, blue light LED, green light LED, yellow light LED, far-red light LED, ultraviolet-A radiation light LED, and ultraviolet-B radiation light LED; W FL and W HPS represent white fluorescent lamps and white high-pressure sodium lamps. ↓, ↑ and — represent that parameter was decreased, increased, and unaffected, respectively.

3.4. Optimizing Harvest Timing and Plant Parts Selections

The absorption and distribution of mineral elements in plants undergo dynamic changes during the various growth stages. K demonstrates high mobility, resulting in fluctuating levels in different plant parts throughout growth [112–114]. For instance, during onion growth stages, K absorption varied significantly: 3.32% at the seedling stage, 71.79% during vigorous growth, and 24.89% during bulb expansion. Distribution patterns also shift, with K primarily found in leaves at the seedling stage, in both bulbs and leaves during vigorous growth, and predominantly in bulbs during bulb expansion [115]. During cultivation, the K content in the roots, leaves, and bulbs initially rose and then decreased, stabilizing over time. After 15 weeks, the K contents per 100 g of FW were 188.9 mg in roots, 158.7 mg in leaves, and 88.3 mg in bulbs [11]. In hydroponic tomato cultivation, the K content in fruits initially increased, peaked at the third truss fruit, and then declined, with the fifth truss fruit exhibiting the lowest content [31]. Consequently, most of the K^+ in leafy vegetables was concentrated in new leaves, allowing old leaves to be harvested and sold as LK vegetable commodities. In contrast, for bulb vegetables or fruits, suitable harvesting times and parts should be determined based on the distribution and variation in K content.

3.5. Application of Exogenous Substances

Certain exogenous substances can mitigate the effects of LK stress on LK fruits and vegetables production, thereby enhancing biomass and nutritional quality. Reported substances include proline [17], gibberellic acid-3 (GA3), jasmonic acid biosynthesis inhibitor (diethyldithiocarbamate; DIECA) [116], indole-3-acetic acid (IAA) [117], 1-naphthaleneacetic acid (NAA) [118], theanine [119], salicylic acid (SA) [120], and silicon [121]. For example, the application of exogenous proline notably increased the proline, soluble sugar, and ascorbic acid levels in LK lettuce. At concentrations of 1 and 1.5 mmol·L^{-1} proline, lettuce leaf superoxide dismutase (SOD) activity increased by 42.5% and 29.5%, respectively. Similarly, peroxidase dismutase (POD) activity increased by 67.0% and 39.3%, whereas shoot FW increased by 37.4% and 31.5%, respectively. Notably, the K content in proline-treated lettuce remained similar to that in untreated samples [17]. Compared to the control group, the supplementation with 0.1 μmol·L^{-1} GA3 increased shoot FW by 1.4 to 2-fold in both the LK-tolerant tomato cultivar, JZ34, and the LK-sensitive cultivar, JZ18. However, the K content in shoots decreased by 23.9% and 21.5% for JZ18 and JZ34, respectively. Similarly, application of 100 $\mu mol \cdot L^{-1}$ DIECA, significantly increased shoot FW by 1.3-fold under LK conditions while slightly reducing the K concentration in shoots [116]. Additionally, Liu [122] demonstrated that the exogenous hormone IAA improved the photosynthetic capacity and antioxidant enzyme activity in sweet potato under LK conditions, thereby reducing the ultrastructural damage in root tip cells. Thus, the appropriate addition of certain exogenous substances can alleviate K deficiency stress in LK fruits and vegetables production.

4. Shortcomings and Prospects of Research on LK Fruits and Vegetables

The LK fruits and vegetables lack standardized industry norms as functional foods for patients with CKD. Currently, there has been no widely accepted K content that defines the LK products. According to a summary of executive conclusions on hyperkalemia management in CKD published by Kidney Disease Improving Global Outcomes, foods with a potassium content of less than 200 $mg \cdot serving^{-1}$ are considered LK foods suitable for CKD patients [123]. A serving of fruit and vegetables is about 150 g and 200 g, respectively [18], so fruits and vegetables with a K content of less than 133 $mg \cdot 100g^{-1}$ and 100 $mg \cdot 100g^{-1}$, respectively, can be considered LK fruits and vegetables. This is in line with the standard for LK fruits and vegetables (less than 100 $mg \cdot 100\ g^{-1}$) currently sold in Japan, but further scientific verification is needed to ensure safety.

The cultivation methods for LK fruits and vegetables are relatively straightforward, often involving a K cut-off at later growth stages. However, this approach may lead to unstable K levels under various cultivation conditions. Although the cultivation primarily focuses on regulating the K content in the nutrient solution, the K levels are also influenced by genotype, growth environment, and other mineral nutrients. Several strategies can be implemented to optimize current cultivation methods. Firstly, K-efficient genotypes have been selected based on previous studies. Secondly, according to the type of fruits and vegetables, scientists can develop specific nutrient solution formula for low-potassium products. And a complete set of fertilization strategies should be developed according to the regulation strategies of irrigation frequency, nutrient solution EC, and pH. Thirdly, environmental controls (temperature and light) have been integrated to boost fruits and vegetables biomass, maximize the dilution effect to enhance KUE, and reduce plant K content. Fourthly, it can enhance the yield and quality by decreasing the K concentration through exogenous substance application, harvest timing, and plant parts selection. Finally, the comprehensive application of these methods can depend on the convenience of cultivation and the economy. In summary, the precise control of K content in fruits and vegetables requires in-depth study to develop a complete cultivation system and establish industry standards.

Currently, the concept of using LK fruits and vegetables for the nutritional treatment of CKD has remained unfamiliar to the public with limited clinical trial data available on their K control ability and safety for CKD patients. CKD nutritional treatment is multifaceted, necessitating the simultaneous consideration of protein, energy, Na, phosphorus, and K intake [18]. Research on LK fruits and vegetables has primarily focused on the K content and lacks comprehensive evaluation standards. Given the importance of fruits and vegetables as staple food items, it is vital to explore and establish accurate sales channels for LK products to meet population demand. Therefore, the research on LK fruits and vegetables should encompass multidisciplinary fields, such as agriculture, medicine, and business, to explore their clinical safety and nutritional treatment methods. Moreover, it is vital to explore potential application scenarios and suitable business models for LK fruits and vegetables to maximize their contribution to the nutritional treatment of CKD.

Author Contributions: Literature search, J.C., L.H., P.L. and C.M.; writing—original draft prepa ration, J.C. and Y.Z.; writing—review and editing, J.Y. and X.D.; writing—supervision, H.Z. an H.W.; financial management, H.J. All authors have read and agreed to the published version o the manuscript.

Funding: This research was funded by the Excellent Team Program of the Shanghai Academy o Agricultural Sciences (grant No. [2022]022).

Data Availability Statement: All data are collected from open sources with detailed descriptions i the cited references.

Conflicts of Interest: The authors declare no conflicts of interest. The funders had no role in th design of the study; in the collection, analyses, or interpretation of data; in the writing of th manuscript; or in the decision to publish the results.

References

1. Borchers, A.; Pieler, T. Programming pluripotent precursor cells derived from *Xenopus* embryos to generate specific tissues an organs. *Genes* **2010**, *1*, 413–426. [CrossRef]
2. Römheld, V.; Kirkby, E.A. Research on potassium in agriculture: Needs and prospects. *Plant Soil* **2010**, *335*, 155–180. [CrossRef]
3. Marschner, P. *Marschner's Mineral Nutrition of Higher Plants*, 3rd ed.; Academic Press: London, UK, 2012.
4. Sustr, M.; Soukup, A.; Tylova, E. Potassium in Root Growth and Development. *Plants* **2019**, *8*, 435. [CrossRef]
5. De La Guardia, M.D.; Benlloch, M. Effects of potassium and gibberellic acid on stem growth of whole sunflower plants. *Physio Plant.* **1980**, *49*, 443–448. [CrossRef]
6. Tränker, M.; Tavakol, E.; Jákli, B. Functioning of potassium and magnesium in photosynthesis, photosynthate translocation an photoprotection. *Physiol. Plant* **2018**, *163*, 414–431. [CrossRef]
7. Greenwood, D.J.; Cleaver, T.J.; Turner, M.K.; Hunt, J.; Niendorf, K.B.; Loquens, S.M.H. Comparison of the effects of potassiur fertilizer on the yield, potassium content and quality of 22 different vegetable and agricultural crops. *J. Agric. Sci.* **1980**, *95* 441–456. [CrossRef]
8. Javaria, S.; Khan, M.Q.; Bakhsh, I. Effect of potassium on chemical and sensory attributes of tomato fruit. *J. Anim. Plant Sci.* **2012** *22*, 1081–1085.
9. Ruiz, R. Effects of different potassium fertilizers on yield, fruit quality and nutritional status of 'Fairlane' nectarine trees and or soil fertility. *Acta Hortic.* **2006**, *721*, 185–190. [CrossRef]
10. Ogawa, A.; Fujita, S.; Toyofuku, K. A cultivation method for lettuce and spinach with high levels of vitamin C using potassiur restriction. *Environ. Control Biol.* **2014**, *52*, 95–99. [CrossRef]
11. Okada, H.; Abedin, T.; Yamamoto, A.; Hayashi, T.; Hosokawa, M. Production of low-potassium onions based on minera absorption patterns during growth and development. *Sci. Hortic.* **2020**, *267*, 109252. [CrossRef]
12. Delanaye, P.; Glassock, R.J.; De Broe, M.E. Epidemiology of chronic kidney disease: Think (at least) twice! *Clin. Kidney J.* **2017**, *1C* 370–374. [CrossRef]
13. Talukder, M.R.; Asaduzzaman, M.; Ueno, M.; Kawaguchi, M.; Yano, S.; Ban, T.; Tanaka, H.; Asao, T. Low potassium conten vegetables research for chronic kidney disease patients in Japan. *Nephrol. Open J.* **2016**, *2*, 1–8. [CrossRef]
14. Kes, P. Hyperkalemia: A potentially lethal clinical condition. *Acta Clin. Croat.* **2001**, *40*, 215–225.
15. Spital, A.; Stems, R.H. Potassium homeostasis in dialysis patients. *Semin. Dial.* **1988**, *1*, 14–20. [CrossRef]
16. Bajwa, S.J.S.; Kwatra, I.S. Nutritional needs and dietary modifications in patients on dialysis and chronic kidney disease. *J. Med Nutr. Nutraceuticals* **2013**, *2*, 46. [CrossRef]
17. Zhang, G.; Yan, Z.; Wang, Y.; Feng, Y.; Yuan, Q. Exogenous proline improve the growth and yield of lettuce with low potassium content. *Sci. Hortic.* **2020**, *271*, 109469. [CrossRef]
18. Cupisti, A.; Kovesdy, C.; D'Alessandro, C.; Kalantar-Zadeh, K. Dietary approach to recurrent or chronic hyperkalaemia in patients with decreased kidney function. *Nutrients* **2018**, *10*, 261. [CrossRef]
19. Renna, M.; Castellino, M.; Leoni, B.; Paradiso, V.; Santamaria, P. Microgreens production with low potassium content for patients with impaired kidney function. *Nutrients* **2018**, *10*, 675. [CrossRef]
20. Tsukagoshi, S.; Hamano, E.; Hohjo, M.; Ikegami, F. Hydroponic production of low-potassium tomato fruit for dialysis patients *Int. J. Veg. Sci.* **2015**, *22*, 451–460. [CrossRef]
21. Ogawa, A.; Taguchi, S.; Kawashima, C. A cultivation method of spinach with a low potassium content for patients on dialysis *Jpn. J. Crop Sci.* **2007**, *76*, 232–237. [CrossRef]
22. Zhang, G.; Johkan, M.; Hohjo, M.; Tsukagoshi, S.; Maruo, T. Plant growth and photosynthesis response to low potassium conditions in three lettuce (*Lactuca sativa*) types. *Hortic. J.* **2017**, *86*, 229–237. [CrossRef]
23. Fuad Mondal, M.; Asaduzzaman, M.; Ueno, M.; Kawaguchi, M.; Yano, S.; Ban, T.; Tanaka, H.; Asao, T. Reduction of potassium (K content in strawberry fruits through KNO_3 management of hydroponics. *Hortic. J.* **2017**, *86*, 26–36. [CrossRef]
24. Asaduzzaman, M.; Talukder, M.R.; Tanaka, H.; Ueno, M.; Kawaguchi, M.; Yano, S.; Ban, T.; Asao, T. Production of low-potassium content melon through hydroponic nutrient management using perlite substrate. *Front. Plant Sci.* **2018**, *9*, 1382. [CrossRef]

25. Tsukagoshi, S.; Aoki, M.; Johkan, M.; Hohjo, M.; Maruo, T. A quantitative management of potassium supply for hydroponic production of low-potassium cherry-type tomato fruit for chronic kidney disease patients. *Horticulturae* **2021**, *7*, 87. [CrossRef]
26. Liu, X. Effect of Low-Potassium Nutrition on the Quality and Potassium Content of Leafy Vegetables. Master's Thesis, South China Agricultural University, Guangzhou, China, 2016.
27. Son, Y.J.; Park, J.E.; Kim, J.; Yoo, G.; Lee, T.S.; Nho, C.W. Production of low potassium kale with increased glucosinolate content from vertical farming as a novel dietary option for renal dysfunction patients. *Food Chem.* **2021**, *339*, 128092. [CrossRef]
28. Xu, H.; Johkan, M.; Tsukagoshi, S.; Maruo, T. Effect of nutrient quantitative management on potassium and sodium concentration in low-potassium lettuce. *Hortic. J.* **2021**, *90*, 154–160. [CrossRef]
29. Flowers, T.J.; Läuchli, A. Sodium versus potassium: Substitution and compartmentation. In *Encyclopedia of Plant Physiology*; Läuchli, A., Pirson, A., Eds.; Springer: Berlin/Heidelberg, Germany, 1983.
30. Liu, G.; Liu, G. The effects of partial replacement of potassium by sodium or calcium in Indica rice (*O. Sativa*). *Acta Agron. Sin.* **1996**, *22*, 313–319.
31. Ogawa, A.; Eguchi, T.; Toyofuku, K. Cultivation methods for leaf vegetables and tomatoes with low potassium content for dialysis patients. *Environ. Control Biol.* **2012**, *50*, 407–414. [CrossRef]
32. Terabayashi, S.; Asaka, T.; Tomatsuri, A.; Date, S.; Fujime, Y. Effect of the limited supply of nitrate and phosphate on nutrient uptake and fruit production of tomato (*Lycopersicon esculentum* Mill.) in hydroponic culture. *Hortic. Res.* **2004**, *3*, 195–200. [CrossRef]
33. Maruo, T.; Takagaki, M.; Shinohara, Y. Critical nutrient concentrations for absorption of some vegetables. *Acta Hortic.* **2004**, *644*, 493–499. [CrossRef]
34. He, J.; See, X.E.; Qin, L.; Choong, T.W. Effects of root-zone temperature on photosynthesis, productivity and nutritional quality of aeroponically grown salad rocket (*Eruca sativa*) vegetable. *Am. J. Plant Sci.* **2016**, *7*, 1993–2005. [CrossRef]
35. Teng, W.; He, X.; Tong, Y. Transgenic approaches for improving use efficiency of nitrogen, phosphorus and potassium in crops. *J. Integr. Agric.* **2017**, *16*, 2657–2673. [CrossRef]
36. Wang, J.; Wang, H.; Zhang, Y.; Zhou, J.; Chen, X. Intraspecific variation in potassium uptake and utilization among sweet potato (*Ipomoea batatas* L.) genotypes. *Field Crops Res.* **2015**, *170*, 76–82. [CrossRef]
37. Yang, H.; Peng, L.; Chen, L.; Zhang, L.; Kan, L.; Shi, Y.; Mei, X.; Malladi, A.; Xu, Y.; Dong, C. Efficient potassium (K) recycling and root carbon (C) metabolism improve K use efficiency in pear rootstock genotypes. *Plant Physiol. Biochem.* **2023**, *196*, 43–54. [CrossRef] [PubMed]
38. Mou, B. Nutrient content of lettuce and its improvement. *Curr. Nutr. Food Sci.* **2009**, *5*, 242–248. [CrossRef]
39. Yang, X.E.; Liu, J.X.; Wang, W.M.; Ye, Z.Q.; Luo, A.C. Potassium internal use efficiency relative to growth vigor, potassium distribution, and carbohydrate allocation in rice genotypes. *J. Plant Nutr.* **2004**, *27*, 837–852. [CrossRef]
40. Wu, J.; Zhang, X.; Li, T.; Yu, H.; Huang, P. Differences in the efficiency of potassium (K) uptake and use in barley varieties. *Agric. Sci. China* **2011**, *10*, 101–108. [CrossRef]
41. Figdore, S.S.; Gabelman, W.H.; Gerloff, G.C. The accumulation and distribution of sodium in tomato strains differing in potassium efficiency when grown under low-K stress. *Plant Soil* **1987**, *99*, 85–92. [CrossRef]
42. Li, Y.; Wang, W.; Wei, K.; Ruan, L.; Wang, L.; Cheng, H.; Zhang, F.; Wu, L.; Bai, P. Differential transcriptomic changes in low-potassium sensitive and low-potassium tolerant tea plant (*Camellia sinensis*) genotypes under potassium deprivation. *Sci. Hortic.* **2019**, *256*, 108570. [CrossRef]
43. George, M.S.; Lu, G.; Zhou, W. Genotypic variation for potassium uptake and utilization efficiency in sweet potato (*Ipomoea batatas* L.). *Field Crops Res.* **2002**, *77*, 7–15. [CrossRef]
44. Jiang, C.; Chen, F.; Gao, X.; Lu, J.; Wan, K.; Nian, F.; Wang, Y. Study on the nutrition characteristics of different K use efficiency cotton genotypes to K deficiency stress. *Agric. Sci. China* **2008**, *7*, 740–745. [CrossRef]
45. Jiang, C.; Xia, Y.; Chen, F.; Lu, J.; Wang, Y. Plant growth, yield components, economic responses, and soil indigenous K uptake of two cotton genotypes with different K-efficiencies. *Agric. Sci. China* **2011**, *10*, 705–713. [CrossRef]
46. Wang, J.; Hou, P.; Zhu, G.; Dong, Y.; Hui, Z.; Ma, H.; Xu, X.; Nin, Y.; Ai, Y.; Zhang, Y. Potassium partitioning and redistribution as a function of K-use efficiency under K deficiency in sweet potato (*Ipomoea batatas* L.). *Field Crops Res.* **2017**, *211*, 147–154. [CrossRef]
47. Liang, S. Stress Screening of Potassium-Efficient Germplasm Resources of Proso Millet (*Panicum miliaceum* L.) and Study on Physiological Mechanism. Master's Thesis, Northwest Agriculture & Forestry University, Yangling, China, 2022.
48. Fan, M.; Bie, Z.; Xie, H.; Zhang, F.; Zhao, S.; Zhang, H. Genotypic variation for potassium efficiency in wild and domesticated watermelons under ample and limited potassium supply. *J. Plant Nutr. Soil Sci.* **2013**, *176*, 466–473. [CrossRef]
49. Zhang, G.; Chen, J.; Eshetu, T.A. Genotypic variation for potassium uptake and utilization efficiency in wheat. *Nutr. Cycl. Agroecosyst.* **1999**, *54*, 41–48.
50. Deng, Z. Screening High Potassium Efficiency Potato Genotypes and Physiological Responses in Different Potassium Level. Master's Thesis, Southwest University, Chongqing, China, 2020.
51. Meng, X.; Liang, T.; Liu, F.; Yu, H.; Hu, L.; Zhang, Y.; Zhou, H.; Zhang, S.; Yin, Q. Potassium absorption and kinetics by different tobacco varieties at different potassium efficiency. *Tob. Sci. Technol.* **2019**, *52*, 1–7. [CrossRef]
52. Díaz-Pérez, J.C. Root zone temperature, plant growth and yield of broccoli [*Brassica oleracea* (Plenck) var. *italica*] as affected by plastic film mulches. *Sci. Hortic.* **2009**, *123*, 156–163. [CrossRef]

53. Yan, Q.; Duan, Z.; Mao, J.; Li, X.; Dong, F. Effects of root-zone temperature and N, P, and K supplies on nutrient uptake of cucumber (*Cucumis sativus* L.) seedlings in hydroponics. *Soil Sci. Plant Nutr.* **2012**, *58*, 707–717. [CrossRef]
54. Zhou, J.; Li, P.; Wang, J.; Fu, W. Growth, photosynthesis, and nutrient uptake at different light intensities and temperatures in lettuce. *HortScience* **2019**, *54*, 1925–1933. [CrossRef]
55. Hood, T.M.; Mills, H.A. Root-zone temperature affects nutrient uptake and growth of snapdragon. *J. Plant Nutr.* **1994**, *17*, 279–29 [CrossRef]
56. Turner, D.W.; Lahav, E. Temperature influences nutrient absorption and uptake rates of bananas grown in controlled environments *Sci. Hortic.* **1985**, *26*, 311–322. [CrossRef]
57. Xia, Z.; Wu, M.; Bai, J.; Zhang, S.; Zhang, G.; Gong, Y.; Yang, Y.; Lu, H. Root zone temperature regulates potassium absorption and photosynthesis in maize (*Zea mays*). *Plant Physiol. Biochem.* **2023**, *198*, 107694. [CrossRef] [PubMed]
58. Kawasaki, Y.; Matsuo, S.; Kanayama, Y.; Kanahama, K. Effect of root-zone heating on root growth and activity, nutrient uptake and fruit yield of tomato at low air temperatures. *J. Jpn. Soc. Hortic. Sci.* **2014**, *83*, 295–301. [CrossRef]
59. Clarke, S.J.; Lamont, K.J.; Pan, H.Y.; Barry, L.A.; Hall, A.; Rogiers, S.Y. Spring root-zone temperature regulates root growth nutrient uptake and shoot growth dynamics in grapevines. *Aust. J. Grape Wine Res.* **2015**, *21*, 479–489. [CrossRef]
60. Giri, A.; Heckathorn, S.; Mishra, S.; Krause, C. Heat stress decreases levels of nutrient-uptake and assimilation proteins in tomato roots. *Plants* **2017**, *6*, 6. [CrossRef]
61. Liu, X.; Huang, B. Root physiological factors involved in cool-season grass response to high soil temperature. *Environ. Exp. Bot.* **2005**, *53*, 233–245. [CrossRef]
62. Tan, L.P.; He, J.; Lee, S.K. Effects of root-zone temperature on the root development and nutrient uptake of *Lactuca sativa* L. "Panama" grown in an aeroponic system in the tropics. *J. Plant Nutr.* **2002**, *25*, 297–314. [CrossRef]
63. Du, Y.C.; Tachibana, S. Effect of supraoptimal root temperature on the growth, root respiration and sugar content of cucumber plants. *Sci. Hortic.* **1994**, *58*, 289–301. [CrossRef]
64. Boatwright, G.O.; Ferguson, H.; Sims, J.R. Soil temperature around the crown node influences early growth, nutrient uptake, and nutrient translocation of spring wheat. *Agron. J.* **1976**, *68*, 227–231. [CrossRef]
65. Matías, J.; Rodríguez, M.J.; Cruz, V.; Calvo, P.; Reguera, M. Heat stress lowers yields, alters nutrient uptake and changes seed quality in quinoa grown under Mediterranean field conditions. *J. Agron. Crop Sci.* **2021**, *207*, 481–491. [CrossRef]
66. Ahammed, G.J.; Chen, Y.; Liu, C.; Yang, Y. Light regulation of potassium in plants. *Plant Physiol. Biochem.* **2022**, *170*, 316–324 [CrossRef] [PubMed]
67. Mankin, K.R.; Fynn, R.P. Nutrient uptake response of New Guinea impatiens to light, temperature, and nutrient solution concentration. *J. Am. Soc. Hortic. Sci.* **1996**, *121*, 826–830. [CrossRef]
68. Walters, K.J.; Currey, C.J. Effects of nutrient solution concentration and daily light integral on growth and nutrient concentration of several basil species in hydroponic production. *HortScience* **2018**, *53*, 1319–1325. [CrossRef]
69. Gerovac, J.R.; Craver, J.K.; Boldt, J.K.; Lopez, R.G. Light intensity and quality from sole-source light-emitting diodes impact growth, morphology, and nutrient content of brassica microgreens. *HortScience* **2016**, *51*, 497–503. [CrossRef]
70. Song, J.; Huang, H.; Hao, Y.; Song, S.; Zhang, Y.; Su, W.; Liu, H. Nutritional quality, mineral and antioxidant content in lettuce affected by interaction of light intensity and nutrient solution concentration. *Sci. Rep.* **2020**, *10*, 2796. [CrossRef]
71. Koukoura, Z.; Kyriazopoulos, A.P.; Parissi, Z.M. Growth characteristics and nutrient content of some herbaceous species under shade and fertilization. *Span. J. Agric. Res. SJAR* **2009**, *7*, 431–438. [CrossRef]
72. Ncise, W.; Daniels, C.W.; Nchu, F. Effects of light intensities and varying watering intervals on growth, tissue nutrient content and antifungal activity of hydroponic cultivated *Tulbaghia violacea* L. under greenhouse conditions. *Heliyon* **2020**, *6*, e03906. [CrossRef]
73. Ma, S.; Lin, J.; Long, Q.; Fan, W.; Cheng, F.; Yang, M. The changes and correlations of mineral nutrition and endogenous hormones in *Erythrophleum fordii* seedlings under different illumination levels. *Hubei Agric. Sci.* **2016**, *55*, 2815–2819. [CrossRef]
74. Xu, J.; Guo, Z.; Jiang, X.; Ahammed, G.J.; Zhou, Y. Light regulation of horticultural crop nutrient uptake and utilization. *Hortic. Plant J.* **2021**, *7*, 367–379. [CrossRef]
75. Nguyen, D.T.P.; Kitayama, M.; Lu, N.; Takagaki, M. Improving secondary metabolite accumulation, mineral content, and growth of coriander (*Coriandrum sativum* L.) by regulating light quality in a plant factory. *J. Hortic. Sci. Biotechnol.* **2019**, *95*, 356–363 [CrossRef]
76. Miao, Y.; Chen, Q.; Qu, M.; Gao, L.; Hou, L. Blue light alleviates 'red light syndrome' by regulating chloroplast ultrastructure photosynthetic traits and nutrient accumulation in cucumber plants. *Sci. Hortic.* **2019**, *257*, 108680. [CrossRef]
77. Chen, X. Effects of Different Light Emitting Diode Sources on Physiological Characteristics and Quality in Chinese Chive. Master's Thesis, Shandong Agricultural University, Tai'an, China, 2012.
78. Yang, X. Effects of Light Quality on the Physiological Characteristics and Quality in Garlic Seedling. Master's Thesis, Shandong Agricultural University, Tai'an, China, 2011.
79. Zheng, X. Effects of Different Light Quality and Nutrient Composition on the Growth and Tipburn Incidence of Lettuce. Master's Thesis, Nanjing Agricultural University, Nanjing, China, 2011.
80. Schroeter-Zakrzewska, A.; Kleiber, T. The effect of light colour and type of lamps on rooting and nutrient status in cuttings of michaelmas daisy. *Bulg. J. Agric. Sci.* **2014**, *20*, 1426–1434.

81. Shin, Y.S.; Lee, M.J.; Lee, E.S.; Ahn, J.H.; Lim, J.H.; Kim, H.J.; Park, H.W.; Um, Y.G.; Park, S.D.; Chai, J.H. Effect of LEDs (light emitting diodes) irradiation on growth and mineral absorption of lettuce (*Lactuca sativa* L. 'Lollo Rosa'). *J. Bio-Environ. Control* **2012**, *21*, 180–185.
82. Shin, Y.S.; Lee, M.J.; Lee, E.S.; Ahn, J.H.; Do, H.W.; Choi, D.W.; Jeong, J.D.; Lee, J.E.; Kim, M.K.; Park, J.U.; et al. Effect of light emitting diodes treatment on growth and mineral contents of lettuce (*Lactuca sativa* L. 'Chung Chi Ma'). *Korean J. Org. Agric.* **2013**, *21*, 659–668. [CrossRef]
83. Pennisi, G.; Blasioli, S.; Cellini, A.; Maia, L.; Crepaldi, A.; Braschi, I.; Spinelli, F.; Nicola, S.; Fernandez, J.A.; Stanghellini, C.; et al. Unraveling the role of red:blue led lights on resource use efficiency and nutritional properties of indoor grown sweet basil. *Front. Plant Sci.* **2019**, *10*, 305. [CrossRef] [PubMed]
84. Chen, X.; Guo, W.; Xue, X.; Beauty, M.M. Effects of LED spectrum combinations on the absorption of mineral elements of hydroponic lettuce. *Spectrosc. Spectr. Anal.* **2014**, *34*, 1394–1397.
85. Wu, J. Effect of Light Quality Ratio on Growth, Quality and Nutrient Absorption of Lettuce. Master's Thesis, South China Agricultural University, Guangzhou, China, 2016.
86. Liu, W.; Zha, L.; Zhang, Y. Growth and nutrient element content of hydroponic lettuce are modified by LED continuous lighting of different intensities and spectral qualities. *Agronomy* **2020**, *10*, 1678. [CrossRef]
87. Tian, J. Effects of Light Conditions and Nutrient Solution Concentration on the Content of Potassium in Lettuce. Master's Thesis, South China Agricultural University, Guangzhou, China, 2018.
88. Kopsell, D.A.; Sams, C.E.; Metallo, R.M.; Waterland, N.L.; Kopsell, D.E. Biomass, carbohydrates, pigments, and mineral elements in kale (*Brassica oleracea* var. *acephala*) microgreens respond to LED blue-light wavelength. *Sci. Hortic.* **2024**, *328*, 112929. [CrossRef]
89. Wei, D. Effects of LED Light Environment on Growth, Quality and Nutrient Absorption of Flowering Chinese Cabbage. Master's Thesis, South China Agricultural University, Guangzhou, China, 2019.
90. Kopsell, D.A.; Sams, C.E.; Barickman, T.C.; Morrow, R.C. Sprouting broccoli accumulate higher concentrations of nutritionally important metabolites under narrow-band light-emitting diode lighting. *J. Am. Soc. Hortic. Sci.* **2014**, *139*, 469–477. [CrossRef]
91. Pinho, P.; Jokinen, K.; Halonen, L. The influence of the LED light spectrum on the growth and nutrient uptake of hydroponically grown lettuce. *Light. Res. Technol.* **2017**, *49*, 866–881. [CrossRef]
92. Cao, K.; Yu, J.; Ye, L.; Zhao, H.; Zou, Z. Optimal LED far-red light intensity in end-of-day promoting tomato growth and development in greenhouse. *Trans. Chin. Soc. Agric. Eng.* **2016**, *32*, 171–176. [CrossRef]
93. Li, Y. Transcriptomic Analysis of the Morphology and Glucosinolate Metabolism in Chinese Kale with Supplemental Far-Red Light. Master's Thesis, South China Agricultural University, Guangzhou, China, 2020.
94. Lőrinc, U.; István, M.; Balázs, V.; Zoltán, P.; Éva, D. Effects of light intensity and spectral composition on the growth and metabolism of spinach (*Spinacia oleracea* L.). *Acta Biol. Plant. Agriensis* **2019**, *7*, 3–18. [CrossRef]
95. Viršilė, A.; Laužikė, K.; Sutulienė, R.; Brazaitytė, A.; Kudirka, G.; Samuolienė, G. Distinct impacts of UV-A Light wavelengths on nutraceutical and mineral contents in green and purple basil cultivated in a controlled environment. *Horticulturae* **2023**, *9*, 1168. [CrossRef]
96. Peng, Q.; Zhou, Q. Effects of enhanced UV-B radiation on the distribution of mineral elements in soybean (*Glycine max*) seedlings. *Chemosphere* **2010**, *78*, 859–863. [CrossRef] [PubMed]
97. Duan, Q.; Zhang, Z.; Chang, P.; Zhang, L.; Wang, J.; He, H. Effects of supplemental light quality and durations of illumination on mineral nutrient absorption and distribution of greenhouse-grown sweet pepper. *J. Nucl. Agric. Sci.* **2020**, *34*, 1814–1825.
98. Li, H.; Liu, H. Effects of supplementary illumination at night on hormones content and nutrient absorption of cucumber seedlings. *Chin. Agric. Sci. Bull.* **2013**, *29*, 74–78.
99. Liu, W.; Zhang, Y.; Zha, L. Effect of LED red and blue continuous lighting before harvest on growth and nutrient absorption of hydroponic lettuce cultivated under different nitrogen forms and light quality. *Spectrosc. Spectr. Anal.* **2020**, *40*, 2215–2221.
100. Mao, J.; Qiu, Q.; Zhang, F.; Li, N.; Hu, Y.; Xue, X. Impact of different photoperiods on the morphological index, quality and absorptive amount to ions of lettuce in fluorescent light source. *North. Hortic.* **2013**, *15*, 24–28.
101. Han, J. Effects of Exogenous Iron on Growth and Fruit Quality of Tomato under Different Photoperiod. Master's Thesis, Shanxi Agricultural University, Taigu, China, 2020.
102. Shao, M.; Liu, W.; Zhou, C.; Wang, Q.; Li, B. Effects of light-dark cycle altered at pre-harvest stage on the growth and nutrient element contents of hydroponic lettuce cultivated under led red and blue light. *China Illum. Eng. J.* **2020**, *31*, 126–130. [CrossRef]
103. Stagnari, F.; Galieni, A.; Cafiero, G.; Pisante, M. Application of photo-selective films to manipulate wavelength of transmitted radiation and photosynthate composition in red beet (*Beta vulgaris* var. *conditiva* Alef.). *J. Sci. Food Agric.* **2013**, *94*, 713–720. [CrossRef]
104. Grygoray, E.E.; Tabalenkova, G.N.; Dalke, I.V.; Golovko, T.K. Mineral nutrition and productivity of the greenhousecucumber crop depending on lighting. *Agrokhimiya* **2015**, *4*, 74–79.
105. Magalhaes, J.R.; Wilcox, G.E. Tomato growth and nutrient uptake patterns as influenced by nitrogen form and light intensity. *J. Plant Nutr.* **2008**, *6*, 941–956. [CrossRef]
106. Lee, M.; Xu, J.; Wang, W.; Rajashekar, C.B. The effect of supplemental blue, red and far-red light on the growth and the nutritional quality of red and green leaf lettuce. *Am. J. Plant Sci.* **2019**, *10*, 2219–2235. [CrossRef]
107. Kim, H.; Yang, T.; Choi, S.; Wang, Y.; Lin, M.; Liceaga, A.M. Supplemental intracanopy far-red radiation to red LED light improves fruit quality attributes of greenhouse tomatoes. *Sci. Hortic.* **2020**, *261*, 108985. [CrossRef]

108. Shen, X.; Li, Z.; Duan, L.; Eneji, A.E.; Li, J. Silicon effects on the partitioning of mineral elements in soybean seedlings unde drought and ultraviolet-B radiation. *J. Plant Nutr.* **2014**, *37*, 828–836. [CrossRef]
109. Yue, M.; Li, Y.; Wang, X. Effects of enhanced ultraviolet-B radiation on plant nutrients and decomposition of spring wheat unde field conditions. *Environ. Exp. Bot.* **1998**, *40*, 187–196. [CrossRef]
110. Yao, X.; Liu, Q. Responses in some growth and mineral elements of mono maple seedlings to enhanced ultraviolet-B and t nitrogen supply. *J. Plant Nutr.* **2009**, *32*, 772–784. [CrossRef]
111. Zhang, Y. Effects of the Combination of Enhanced UV-B Radiation and Cd^{2+} on Photosynthesis and Metal Elements Absorptior in Mung Bean Seedlings at the Initial Stage of Growth. Master's Thesis, Shaanxi Normal University, Xi'an, China, 2005.
112. Marschner, H. Functions of mineral nutrients: Micronutrients. In *Mineral Nutrition of Higher Plants*; Marschner, R., Ed.; Academi Press: London, UK, 1995; pp. 313–404.
113. Karley, A.J.; White, P.J. Moving cationic minerals to edible tissues: Potassium, magnesium, calcium. *Curr. Opin. Plant Biol.* **2009** *12*, 291–298. [CrossRef] [PubMed]
114. Zörb, C.; Senbayram, M.; Peiter, E. Potassium in agriculture—Status and perspectives. *J. Plant Physiol.* **2014**, *171*, 656–669 [CrossRef]
115. Zhao, K.; Xu, K.; Xu, N.; Wang, Y. Absorption and distribution of nitrogen, phosphorus and potassium in onion. *Plant J. Fertil. Sc* **2009**, *15*, 241–246.
116. Wang, X.; Luo, J.F.; Liu, R.; Liu, X.; Jiang, J. Gibberellins regulate root growth by antagonizing the jasmonate pathway in tomat plants in response to potassium deficiency. *Sci. Hortic.* **2023**, *309*, 111693. [CrossRef]
117. Liu, M.; Yu, Y.; Jin, R.; Zhao, P.; Zhang, Q.; Zhu, X.; Wang, J.; Tang, Z. Exogenous IAA enhances low potassium tolerance o sweetpotato by regulating root response strategy. *Arch. Agron. Soil Sci.* **2023**, *70*, 1–15. [CrossRef]
118. Zhou, D.; Zhang, Y.; Dong, Q.; Wang, K.; Zhang, H.; Du, Q.; Wang, J.; Wang, X.; Yu, H.; Zhao, X. Integrated transcriptomic anc metabolomic analysis of exogenous NAA effects on maize seedling root systems under potassium deficiency. *Int. J. Mol. Sci.* **2024** *25*, 3366. [CrossRef] [PubMed]
119. Liu, Y.; Shan, D.; Liu, S.; You, B.; Guo, J.; Chen, X. Protective role of theanine on tobacco seedlings under low potassium stress. *J. Anhui Agric. Univ.* **2016**, *43*, 282–287. [CrossRef]
120. Guo, M.; Zhu, Y.; Huang, W.; Si, D.; Li, H. Effects of exogenous salicylic acid on the growth of *Zinnia elegans* seedlings under low potassium stress. *J. Henan Agric. Sci.* **2019**, *48*, 146–150. [CrossRef]
121. Yang, R.; Guo, S.; Yang, X.; Zhang, Y.; Shi, Y. Effects of silicon on growth and physiological characteristics of tomato seedlings under low potassium stress. *Shandong Agric. Sci.* **2022**, *54*, 55–63. [CrossRef]
122. Liu, M.; Zhang, A.; Chen, X.; Sun, H.; Jin, R.; Li, H.; Tang, Z. Regulation of leaf and root physiology of sweet potato (*Ipomoea batatas* [L.] Lam.) with exogenous hormones under potassium deficiency stress. *Arch. Agron. Soil Sci.* **2019**, *66*, 600–613. [CrossRef]
123. Clase, C.M.; Carrero, J.J.; Ellison, D.H.; Grams, M.E.; Hemmelgarn, B.R.; Jardine, M.J.; Kovesdy, C.P.; Kline, G.A.; Lindner, G. Obrador, G.T.; et al. Potassium homeostasis and management of dyskalemia in kidney diseases: Conclusions from a Kidney Disease: Improving Global Outcomes (KDIGO) Controversies Conference. *Kidney Int.* **2020**, *97*, 42–61. [CrossRef]

plants MDPI

Article

Ascorbic Acid Improves Tomato Salt Tolerance by Regulating Ion Homeostasis and Proline Synthesis

Xianjun Chen [1,2,†], Hongwei Han [1,3,†], Yundan Cong [1,†], Xuezhen Li [1], Wenbo Zhang [1], Jinxia Cui [1], Wei Xu [1], Shengqun Pang [1,*] and Huiying Liu [1,*]

1 Key Laboratory of Special Fruits and Vegetables Cultivation Physiology and Germplasm Resources Utilization of Xinjiang Production and Contruction Crops, Department of Horticulture, Agricultural College, Shihezi University, Shihezi 832003, China; chenxianjun_0805@163.com (X.C.); xjnykx-h@xaas.ac.cn (H.H.); yundan_cong@163.com (Y.C.); lixuezhen492@163.com (X.L.); zhangwb626@163.com (W.Z.); jinxiacui77@163.com (J.C.); xuwei0412@shzu.edu.cn (W.X.)

2 Key Laboratory of Molecular Breeding and Variety Creation of Horticultural Plants for Mountain Features in Guizhou Province, School of Life and Health Science, Kaili University, Kaili 556011, China

3 Key Laboratory of Horticulture Crop Genomics and Genetic Improvement in Xinjiang, Institute of Horticultural Crops, Xinjiang Academy of Agricultural Sciences, Urumqi 830000, China

* Correspondence: pangshqok@shzu.edu.cn (S.P.); liuhy_bce@shzu.edu.cn (H.L.)

† These authors contributed equally to this work.

Abstract: In this study, processing tomato (*Solanum lycopersicum* L.) 'Ligeer 87-5' was hydroponically cultivated under 100 mM NaCl to simulate salt stress. To investigate the impacts on ion homeostasis, osmotic regulation, and redox status in tomato seedlings, different endogenous levels of ascorbic acid (AsA) were established through the foliar application of 0.5 mM AsA (NA treatment), 0.25 mM lycorine (LYC, an inhibitor of AsA synthesis; NL treatment), and a combination of LYC and AsA (NLA treatment). The results demonstrated that exogenous AsA significantly increased the activities and gene expressions of key enzymes (L-galactono-1,4-lactone dehydrogenase (GalLDH) and L-galactose dehydrogenase (GalDH)) involved in AsA synthesis in tomato seedling leaves under NaCl stress and NL treatment, thereby increasing cellular AsA content to maintain its redox status in a reduced state. Additionally, exogenous AsA regulated multiple ion transporters via the SOS pathway and increased the selective absorption of K^+, Ca^{2+}, and Mg^{2+} in the aerial parts, reconstructing ion homeostasis in cells, thereby alleviating ion imbalance caused by salt stress. Exogenous AsA also increased proline dehydrogenase (ProDH) activity and gene expression, while inhibiting the activity and transcription levels of Δ1-pyrroline-5-carboxylate synthetase (P5CS) and ornithine-δ-aminotransferase (OAT), thereby reducing excessive proline content in the leaves and alleviating osmotic stress. LYC exacerbated ion imbalance and osmotic stress caused by salt stress, which could be significantly reversed by AsA application. Therefore, exogenous AsA application increased endogenous AsA levels, reestablished ion homeostasis, maintained osmotic balance, effectively alleviated the inhibitory effect of salt stress on tomato seedling growth, and enhanced their salt tolerance.

Keywords: ascorbic acid; ion homeostasis; osmotic balance; salt stress; tomato

check for updates

Citation: Chen, X.; Han, H.; Cong, Y.; Li, X.; Zhang, W.; Cui, J.; Xu, W.; Pang, S.; Liu, H. Ascorbic Acid Improves Tomato Salt Tolerance by Regulating Ion Homeostasis and Proline Synthesis. *Plants* **2024**, *13*, 1672. https://doi.org/10.3390/plants13121672

Academic Editor: Daniela Businelli

Received: 5 June 2024
Revised: 12 June 2024
Accepted: 13 June 2024
Published: 17 June 2024

1. Introduction

Soil salinization poses a significant challenge to agricultural production, impeding crop growth and limiting yield formation [1,2]. Tomato (*Solanum lycopersicum* L.), widely cultivated worldwide, is crucial for global agricultural production and trade. Xinjiang, China's largest processing tomato production base, is severely threatened by soil salinization. Salt stress reduces the water potential of plant roots, leading to physiological drought and inducing osmotic stress [3]. Excess salts absorbed by roots enter the aboveground parts through transpiration, causing the accumulation of excessive Na^+ and Cl^- in leaves, which compete with ions such as Ca^{2+}, K^+, and Mg^{2+} and disrupt ionic homeostasis [4].

Osmotic stress and ionic imbalance induce the production of reactive oxygen species (ROS), which in turn causes oxidative stress, lipid peroxidation, and irreparable damage to cellular membranes, ultimately impacting plant growth and development [5]. Consequently, reducing Na^+ accumulation, restoring ionic homeostasis, and maintaining osmotic balance and redox status are crucial for mitigating the adverse effects of salt stress on plants.

During the course of evolution, plants have developed a mechanism to maintain low levels of Na^+ by actively removing Na^+ from the cytoplasm. Under salt stress, Na^+ accumulation occurs, and plant cells sequester Na^+ into vacuoles via Na^+/H^+ antiporters, thereby reducing ion damage [6]. The salt overly sensitive (SOS) signaling pathway regulates ion homeostasis by modulating the activity of Na^+/H^+ antiporters in response to salt stress [7]). In addition to inorganic ions, proline (Pro) plays a crucial role as an osmotic regulator in plants under stress conditions. The increase in proline content in plants under salt stress is tightly regulated by enzymes involved in proline metabolism. Proline biosynthesis in plants primarily occurs through the glutamate (Glu) pathway and the ornithine (Orn) pathway, with the former being predominantly catalyzed by Δ1-pyrroline-5-carboxylate synthetase (P5CS) and the latter mainly regulated by ornithine δ-aminotransferase (OAT) [8]. Most studies indicate that the P5CS pathway is the primary route for proline accumulation during stress [9], while OAT plays a key role in regulating plant cell redox homeostasis by modulating proline metabolism under stress conditions [10].

In recent years, the use of exogenous substances to alleviate salt stress damage in plants has emerged as an effective strategy to enhance plant salt tolerance [11–13]. L-ascorbic acid (AsA), acting as an electron donor in redox reactions and an antioxidant, plays crucial roles in plant growth, development, and stress responses [14,15]. AsA, located in the cytoplasm and extracellular space, can directly perceive environmental stressors, thereby regulating antioxidant defense and redox-sensitive signal transduction pathways [16,17]. Moreover, AsA has been shown to effectively enhance plant stress resilience. The application of exogenous AsA can shield lipids and proteins from oxidative damage induced by salt or drought stress [18,19]. Furthermore, numerous studies have underscored the critical role of suitable concentrations of exogenous AsA in ameliorating damage caused by various abiotic stresses, including salt [20], low temperature [21], heavy metals [22,23] and drought [24]. Current research on alleviating salt stress with exogenous AsA primarily focuses on antioxidant defense mechanisms and the mitigation of photoinhibition. For instance, exogenous AsA enhances the activities of antioxidant enzymes in wheat [25], tomato [26], strawberry [20], and cowpea [27] under salt stress, thereby counteracting the adverse effects of salt stress on plant growth. However, research on the regulation of ion homeostasis, selective absorption, and transport, the regulatory mechanism of Na^+ regulation, and the synthesis and metabolism regulation of proline under salt stress by exogenous AsA remains limited.

Previous studies have demonstrated that exogenous AsA can increase endogenous AsA content [28]. However, it remains unclear whether exogenous AsA increases endogenous AsA levels by regulating the expression and activity of key enzymes involved in AsA biosynthesis. Therefore, in this study, processing tomatoes were used as the experimental material, and exogenous AsA and LYC (lycorine, an inhibitor of AsA synthesis) were applied to create different AsA levels. The study aimed to elucidate the mechanism by which exogenous AsA enhances the salt tolerance of processing tomato seedlings from the perspectives of ion homeostasis and osmotic stress. The results demonstrate that exogenous AsA maintained a high level of endogenous AsA and the redox pool by altering the activities of enzymes related to endogenous AsA synthesis and the expression levels of related genes. It induced the SOS pathway to regulate multiple ion transporters, adjusting intracellular ion homeostasis. Moreover, it enhanced the selective transport of K^+, Ca^{2+}, and Mg^{2+} in the aerial parts, facilitated the efflux and compartmentalization of Na^+ and Cl^- and alleviated ion imbalance caused by salt stress. Additionally, exogenous AsA regulated osmotic adjustment by enhancing Pro degradation and inhibiting its synthesis, thereby mitigating the damage of salt stress to plants. Ultimately, these findings suggest that exogenous AsA effectively enhances root vitality and growth characteristics of processing tomato seedlings under salt stress.

2. Materials and Methods

2.1. Plant Materials and Treatment Conditions

The experiment was conducted at Shihezi University's solar greenhouse using the processing tomato cultivar 'Ligeer 87-5′. Seeds were germinated in plugs with a charcoal and vermiculite mixture (2:1 v/v). After producing two true leaves, uniform seedlings were transferred to black plastic buckets with foam covers for hydroponics, each containing 10 L of Hoagland nutrient solution (pH = 6.2) diluted with deionized water. After a 7-day pre-cultivation period, seedlings were subjected to various treatments by adding 100 mM NaCl to the nutrient solution for salt stress. Daily leaf sprays of 0.5 mM ascorbic acid (AsA) and 0.25 mM lycorine (LYC), an inhibitor of the key AsA synthesis enzyme L-galactono-1, 4-lactone dehydrogenase (GalLDH), were applied. AsA and LYC were sourced from Sigma (St. Louis, MO, USA) and Yuan Ye (China), respectively. Concentrations were based on prior screening experiments. The five treatments were (1) control: distilled water; (2) NaCl: 100 mM NaCl + distilled water; (3) NA: 100 mM NaCl + 0.5 mM AsA; (4) NL: 100 mM NaCl + 0.25 mM LYC; and (5) NLA: 100 mM NaCl + 0.25 mM LYC + 0.5 mM AsA. A completely randomized block design with four replications per treatment and five plants per replication was used. The nutrient solution was oxygenated throughout the experiment, with sampling on the third day of treatment [28].

2.2. Determination of Growth Indicators

Aboveground and belowground relative growth rates (RGRs) were calculated following the method described by Van [29]. Root activity was determined using the triphenyltetrazolium chloride (TTC) method, as outlined by Li [30].

2.3. Ion Content and Transcriptional Expression Assay of Key Genes of SOS Signaling Pathway

K^+, Ca^{2+}, Na^+, and Mg^{2+} ions were quantified using inductively coupled plasma emission spectrometry (ICP-OES, Agilent, Santa Clara, CA, USA). Cl^- content was determined following Nazar's method [31]. The ion-selective transport ratios [$S_{Na, X}$ = leaf (X/Na^+)/root (X/Na^+)] were calculated according to Epstein's method [32], with X representing Ca^{2+}, K^+, and Mg^{2+}, respectively.

Gene expression levels of salt overly sensitive 1,2,3 (*SOS1, SOS2, SOS3*), Na^+/H^+ antiporter 1, 2, 3 (*NHX1, NHX2, NHX3*), high-affinity potassium transporter protein (*HKT1;2*), pyrophosphate-energized vacuolar membrane proton pump (*VP1*), and chloride channel (*CLC*) were determined following Livak et al. [33], with the primers used detailed in Table 1.

Table 1. Quantitative real-time PCR sequences.

Gene	Primer	Sequence (5′ to 3′)
Actin (NM_001323002.1)	FORWARD	TGGTCGGAATGGGAAAG
	REVERSE	CTCAGTCAGGAGAACAGGGT
SOS1 (AJ717346.1)	FORWARD	GCTGATGTCTCTGGTGTCTTGACTG
	REVERSE	TTGATGACTCTCGCCCTTGAAAGC
SOS2 (NM_001247281.2)	FORWARD	TATTTCCCGCCAACCTGCTAAAGTC
	REVERSE	GACCAGCCCTATTTGCCGTTACC
SOS3 (AJ717347.1)	FORWARD	TATTCCACCCAAATGCACCAGTAGC
	REVERSE	CATTCAGCAGCGCCAAAACCATC
NHX1 (NM_001246987.1)	FORWARD	CTTGGTCTGGTTCTGGTTGGAAGG
	REVERSE	AGCCCACCATATCGTGACCTGTAG
NHX2 (NM_001328634.1)	FORWARD	TCACTGCTACCACTGCCATTGTTG
	REVERSE	ACCATCACCCACAACTTCCAAAGC
NHX3 (NM_001247326.2)	FORWARD	TGGTTGGAAGGGCAGCATTTGTC
	REVERSE	TGAAACAGCACCTCGCATAAGTCC

Table 1. *Cont.*

Gene	Primer	Sequence (5′ to 3′)
HKT1;2 (NM_001302904.1)	FORWARD	CCTACCGTCTTTTCGTCCTCA
	REVERSE	GCTTCCCCACCAAGAAACATC
VP1 (NM_001278976.2)	FORWARD	GATGGTTGAGGAAGTGCGTAGGC
	REVERSE	CACAGGTGGCATAGTCAGGCTTG
CLC (NM_001247096.2)	FORWARD	CGTCGCCTTCGCCTTCTAATCG
	REVERSE	CAACAAGCAACATCGCCCATTCC
P5CS (NM_001246978.2)	FORWARD	TGGAAGATTAGGAGCCCTCTGTGAG
	REVERSE	CTAAGCCGCTGACGACCAACAC
OAT (NM_001247674.3)	FORWARD	GGCTCTCATTGTCTCGTGCTGTG
	REVERSE	GGGCAACCGAATCTCCAAAATCAAC
ProDH (NM_001347105.1)	FORWARD	CCACCACCACGACCATCACAAC
	REVERSE	CATTACCCACATGCCCAAATCAACC
AAO (NM_001247900.2)	FORWARD	ACAAGCAGGACTACAAGGAATGATG
	REVERSE	AGGCAATGAAGCAAGACCAGTTG
GalDH (XM_004230609.4)	FORWARD	CAACGACTGGAATGGACGAAGAAG
	REVERSE	AACAGGAGATCACAATTCACAAGACC
GalLDH (NM_001247674.3)	FORWARD	GTTGAGAGGCAGGAGCTTGTAGAAC
	REVERSE	TGTCACAACCACAACGGCATCAG

Note: *Actin*: Actin gene; *SOS1*: Salt overly sensitive 1 gene; *SOS2*: Salt overly sensitive 2 gene; *SOS3*: Salt overly sensitive 3 gene; *NHX1*: Na^+/H^+ antiporter 1 gene; *NHX2*: Na^+/H^+ antiporter 2 gene; *NHX3*: Na^+/H^+ antiporter 3 gene; *HKT1;2*: high-affinity potassium transporter protein gene; *VP1*: pyrophosphate-energized vacuolar membrane proton pump gene; *CLC*: chloride channel protein gene; *P5CS*: Δ1-pyrroline-5-carboxylate synthase gene; *OAT*: ornithine-δ-aminotransferase gene; *ProDH*: proline dehydrogenase gene; *AAO*: ascorbate oxidase gene; *GalDH*: L-galactose dehydrogenase gene; *GalLDH*: L-galactono-1, 4-lactone dehydrogenase gene.

2.4. Proline (Pro) Content and Its Anabolic Key Enzyme Activities and Gene Expression Assays

Proline (Pro) content was quantified using the acid ninhydrin colorimetry method described by de Freitas [34]. The activity of Δ1-pyrroline-5-carboxylate synthase (P5CS) was assessed following the protocol of Song et al. [35], while the activity of ornithine-δ-aminotransferase (OAT) was measured according to Kim et al. [36]. Proline dehydrogenase (ProDH) activity was determined using the method outlined by Lutts [37]. A gene expression analysis of key enzymes involved in Pro synthesis and metabolism was performed using qRT-PCR, with the primer sequences provided in Table 1.

2.5. Ascorbic Acid (AsA) Content and Its Anabolic Key Enzyme Activities and Gene Expression Assays

Ascorbic acid (AsA) and dehydroascorbic acid (DHA) levels were measured following the protocol outlined by Jiang et al. [38]; l-galactose dehydrogenase (GalDH) activity was determined using the method described by Gatzek [39]; L-galactono-1, 4-lactone dehydrogenase (GalLDH) activity was assessed according to Ôba [40]; and ascorbate oxidase (AAO) activity was determined following the procedure by Esaka [41]. The gene expression of *GalDH*, *GalLDH*, and *AAO* was analyzed using qRT-PCR, with primer sequences detailed in Table 1.

2.6. Gene Expression Analysis

The total RNA from tomato leaves was extracted using the Trizol method and reverse transcribed into cDNA using the Hyper ScriptTM III RT SuperMix (EnzyArtisan Biotech, Shanghai, China), following the manufacturer's instructions. qPCR amplification was performed in real-time using 2 × S6 Universal SYBR qPCR Mix (enzyme Biotech, China). Each sample was run in triplicate, and each gene was analyzed with three biological and technical replicates. The relative gene expression was calculated using the $2^{-\Delta\Delta Ct}$ method. The qRT-PCR amplification primers are listed in Table 1. The tomato actin gene served as an internal control [42,43].

2.7. Statistical Analysis

Data were processed and statistically analyzed using Microsoft Excel 2020 and SPSS 19.0. Graphs were generated using Origin 2021 software. One-way analysis of variance (ANOVA) followed by Duncan's multiple range test was utilized to ascertain the significance of differences between treatments ($p < 0.05$). The results are expressed as mean ± standard deviation.

3. Results

3.1. Exogenous AsA Promotes the Growth of Tomato Seedlings under Salt Stress

Figure 1 demonstrates that 100 mM NaCl stress significantly hindered the growth of tomato seedlings, as indicated by a notable decrease in the relative growth rate of both aerial and underground parts, along with the root activity of tomato seedlings compared to the control. However, the application of exogenous AsA mitigated the adverse effects of NaCl stress on the relative growth rate of both aerial and underground parts, as well as the root activity of tomato seedlings, resulting in significant increases of 186.1%, 141.9%, and 66.9%, respectively. In contrast, LYC application (NL treatment) exacerbated the inhibitory effect of NaCl stress on tomato seedling growth. The NLA treatment significantly reversed the aforementioned indices compared to the NL treatment. These results indicate that AsA can promote the growth of tomato seedlings under salt stress.

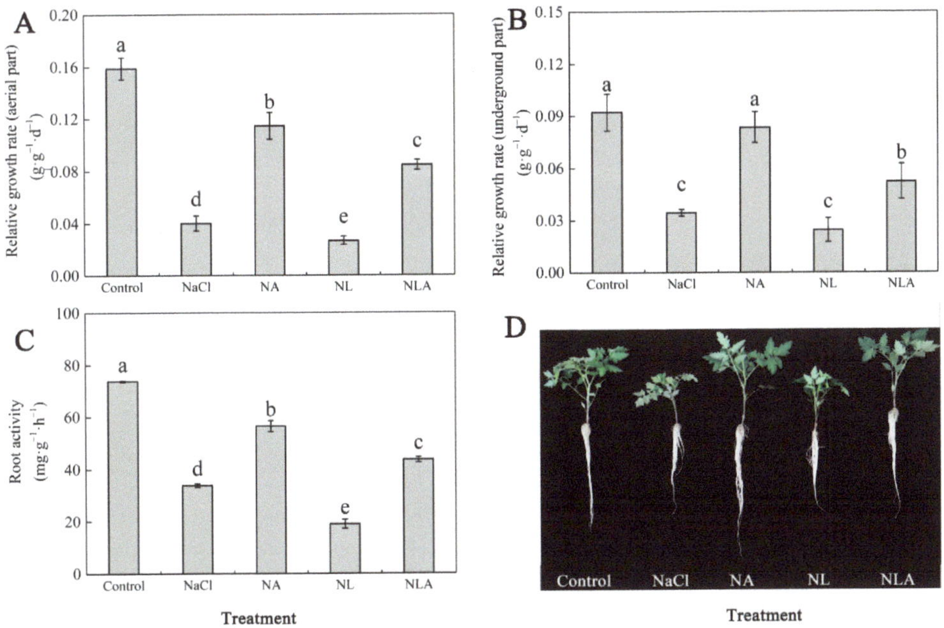

Figure 1. Values of the relative growth rate (aerial part) (**A**), relative growth rate (underground part) (**B**), root activity (**C**), and photographs of tomato seedlings on the third day after different treatments (**D**) in the leaves of salt-stressed tomato seedlings with or without exogenous reduced ascorbic acid (AsA) and Lycorine (LYC, AsA synthesis inhibitor) spraying. All measurements were performed on leaves at 3 d after treatment. Error bars represent SD (n = 4). Different letters indicate significance differences among treatments ($p < 0.05$). Control: no NaCl and no AsA and no LYC; NaCl: 100 mM NaCl; NA: 100 mM NaCl + 0.5 mM AsA; NL: 100 mM NaCl + 0.25 mM LYC; NLA: 100 mM NaCl + 0.25 mM LYC + 0.5 mM AsA.

3.2. Exogenous AsA Increases the Content of Endogenous AsA of Tomato Seedlings under Salt Stress

As depicted in Figure 2, exposure to NaCl stress significantly reduced the endogenou AsA content, total AsA content, and the AsA/DHA ratio, while significantly increasing th DHA content compared to the control. The application of exogenous AsA (NA treatmen resulted in a significant increase in AsA content by 172.9%, total AsA content by 53.1% and the AsA/DHA ratio by 338.7%, while significantly decreasing the DHA content b 37.8% compared to the NaCl treatment. Conversely, the NL treatment significantly reduce the AsA and total AsA content, as well as the AsA/DHA ratio, compared to the NaC treatment, with a significant increase in DHA content in tomato leaves. The NLA treatmen significantly decreased the DHA content and significantly increased the AsA and tota AsA content, as well as the AsA/DHA ratio, compared to the NL treatment. These result indicate that exogenous AsA can increase the content of endogenous AsA and maintain high redox pool of AsA.

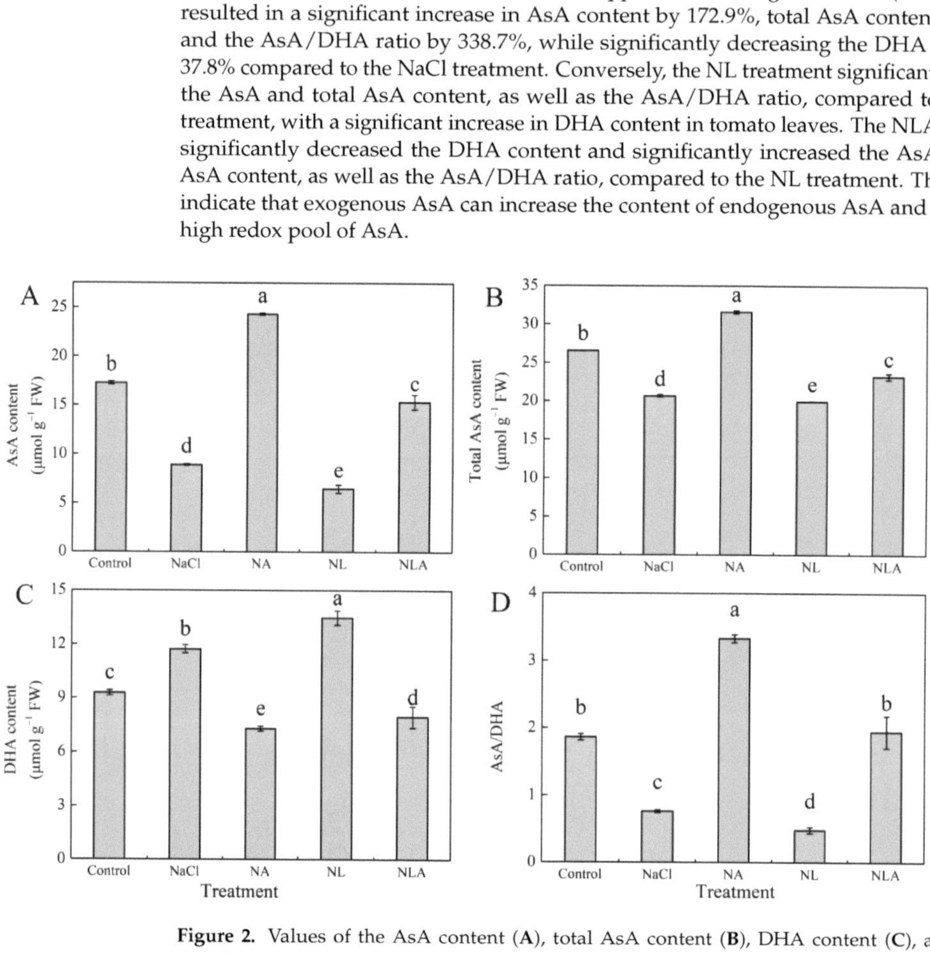

Figure 2. Values of the AsA content (**A**), total AsA content (**B**), DHA content (**C**), and ratio o AsA/DHA (**D**) in the leaves of salt-stressed tomato seedlings with or without exogenous reduce ascorbic acid (AsA) and Lycorine (LYC, AsA synthesis inhibitor) spraying. All measurements were performed on leaves at 3 d after treatment. Error bars represent SD ($n = 4$). Different letters indicate significance differences among treatments ($p < 0.05$). Control: no NaCl and no AsA and no LYC NaCl: 100 mM NaCl; NA: 100 mM NaCl + 0.5 mM AsA; NL: 100 mM NaCl + 0.25 mM LYC; NLA 100 mM NaCl + 0.25 mM LYC + 0.5 mM AsA.

3.3. Exogenous AsA Affects Key Enzyme Activities and the Gene Expression of AsA Anabolism of Tomato Seedlings under Salt Stress

As shown in Figure 3, the activities of GalDH and GalLDH in tomato seedling leaves under NaCl stress were significantly reduced, while AAO activity was significantly increased compared to the control, exhibiting similar trends in *AAO*, *GalDH*, and *GalLDH* gene expression. The exogenous application of AsA significantly increased the enzyme activities of GalDH and GalLDH and their gene expression, while significantly decreasing AAO activity and gene expression levels compared to the NaCl treatment. Conversely, the exogenous application of LYC significantly decreased GalDH and GalLDH activities and *GalDH* gene expression levels compared to the NaCl treatment, had no significant effect on AAO activity, but significantly upregulated *AAO* gene expression levels throughout the treatment. The application of AsA on top of the NL treatment significantly reversed these indices, increasing GalDH and GalLDH activities by 103.1% and 129.3%, respectively, and significantly decreasing AAO activity by 90.7%. These results indicate that exogenous AsA can significantly affect the activity of key enzymes and the gene expression of endogenous AsA anabolic metabolism.

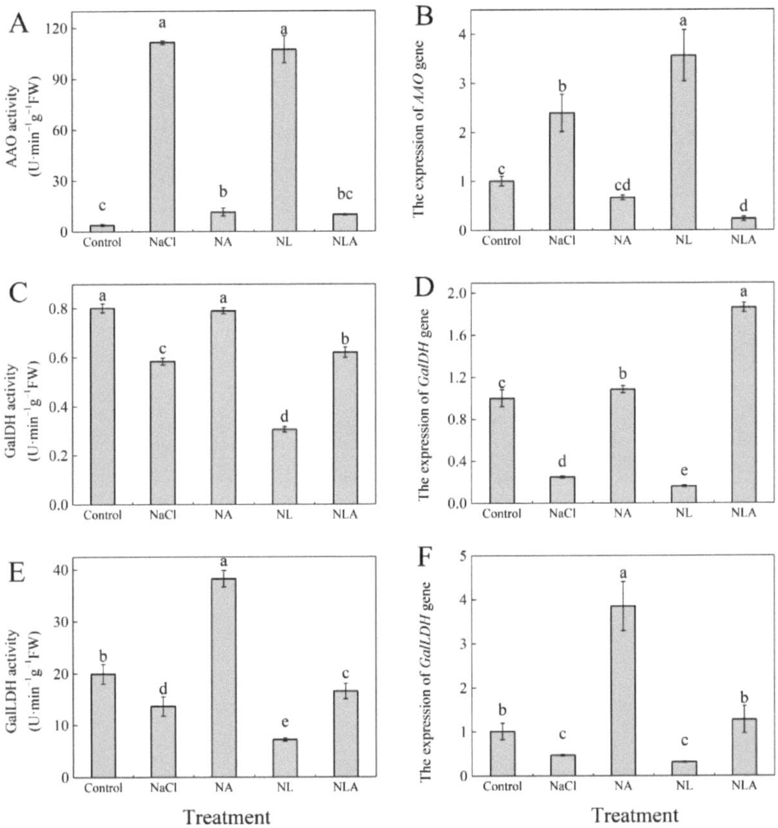

Figure 3. Values of the AAO activity (**A**), *AAO* gene expression (**B**), GalDH activity (**C**), *GalDH* gene expression (**D**), GalLDH activity (**E**), and *GalLDH* gene expression (**F**) in the leaves of salt-stressed tomato seedlings with or without exogenous reduced ascorbic acid (AsA) and Lycorine (LYC, AsA synthesis inhibitor) spraying. All measurements were performed on leaves at 3 d after treatment. Error bars represent SD ($n = 4$). Different letters indicate significance differences among treatments ($p < 0.05$). Control: no NaCl and no AsA and no LYC; NaCl: 100 mM NaCl; NA: 100 mM NaCl + 0.5 mM AsA; NL: 100 mM NaCl + 0.25 mM LYC; NLA: 100 mM NaCl + 0.25 mM LYC + 0.5 mM AsA.

3.4. Exogenous AsA Alleviates the Ionic Imbalance of Tomato Seedlings under Salt Stress

Compared to the control, the NaCl treatment significantly increased the Na^+ and Cl^- contents as well as Na^+/K^+, Na^+/Ca^{2+}, and Na^+/Mg^{2+} ratios in tomato leaves and roots, and significantly decreased K^+, Ca^{2+}, and Mg^{2+} contents, as shown in Figure 4. Treatment with exogenous AsA under salt stress significantly reduced the Na^+ and Cl^- contents in both leaves and roots by 2.6% and 14.1%, and 21.6% and 16.6%, respectively, compared to the NaCl treatment. Furthermore, it significantly increased the K^+, Ca^{2+}, and Mg^{2+} contents by 10.5% and 13.6%, 36.4% and 52.6%, and 15.8% and 7.8%, respectively. Additionally, the ratios of Na^+/K^+, Na^+/Ca^{2+}, and Na^+/Mg^{2+} showed a significant decrease. The NL treatment led to an accumulation of Na^+ and Cl^- and a significant reduction in K^+, Ca^{2+}, and Mg^{2+} contents in both leaves and roots of tomato seedlings under NaCl treatment, further disrupting the ionic homeostasis (resulting in significantly increased Na^+/K^+, Na^+/Ca^{2+}, and Na^+/Mg^{2+} ratios). However, the application of AsA significantly mitigated the negative effects of LYC on the leaves and roots of tomato seedlings under salt stress. The above indicates that exogenous AsA can reduce the ion imbalance caused by salt stress in tomato seedlings.

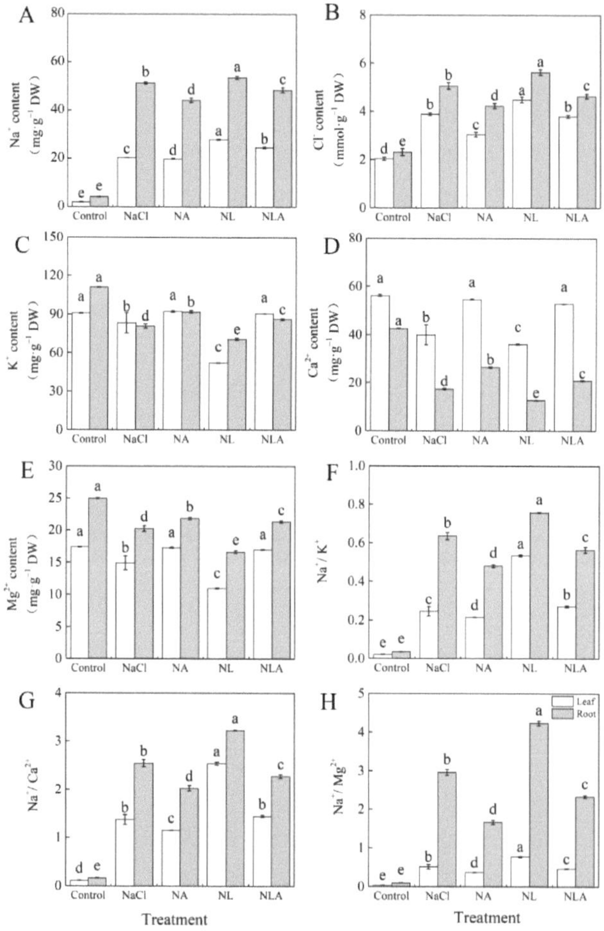

Figure 4. Values of the Na^+ content (**A**), Cl^- content (**B**), K^+ content (**C**), Ca^{2+} content (**D**), Mg^{2+} content (**E**), Na^+/K^+ (**F**), Na^+/Ca^{2+} (**G**) and Na^+/Mg^{2+} (**H**) in the leaves and roots of salt-stressed tomato seedlings with or without exogenous reduced ascorbic acid (AsA) and Lycorine (LYC, AsA synthesis inhibitor) spraying. All measurements were performed on leaves at 3 d after treatment. Error bars represent SD ($n = 4$). Different letters indicate significance differences among treatments ($p < 0.05$). Control: no NaCl and no AsA and no LYC; NaCl: 100 mM NaCl; NA: 100 mM NaCl + 0.5 mM AsA; NL: 100 mM NaCl + 0.25 mM LYC; NLA: 100 mM NaCl + 0.25 mM LYC + 0.5 mM AsA.

3.5. Exogenous AsA Affects the Ion Selective Absorption and Transportation Capacity of Tomato Seedlings under Salt Stress

As shown in Figure 5, compared to the control, salt stress significantly increased $S_{K, Na}$, $S_{Mg, Na}$, and $S_{Ca, Na}$ by 52.6%, 28.4% and 110.5%, respectively. The exogenous spraying of AsA significantly decreased $S_{K, Na}$, $S_{Mg, Na}$, and $S_{Ca, Na}$ in tomato seedlings under salt stress. Conversely, the exogenous spraying of LYC significantly decreased $S_{K, Na}$ and $S_{Mg, Na}$ but had no significant effect on $S_{Ca, Na}$ under salt stress. Compared to the NL treatment, the NLA treatment significantly increased $S_{K, Na}$ and $S_{Mg, Na}$, while significantly decreasing $S_{Ca, Na}$. These results indicate that exogenous AsA can significantly affect the ion selective transport ratio of tomato under salt stress.

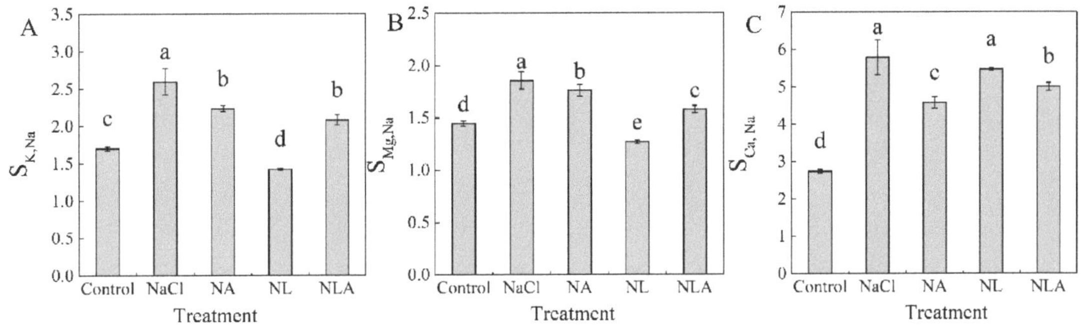

Figure 5. Values of $S_{K, Na}$ (**A**), $S_{Mg, Na}$ (**B**), and $S_{Ca, Na}$ (**C**) in leaves of salt-stressed tomato seedlings with or without exogenous reduced ascorbic acid (AsA) and Lycorine (LYC, AsA synthesis inhibitor) spraying. All measurements were performed on leaves at 3 d after treatment. Error bars represent SD ($n = 4$). Different letters indicate significance differences among treatments ($p < 0.05$). Control: no NaCl and no AsA and no LYC; NaCl: 100 mM NaCl; NA: 100 mM NaCl + 0.5 mM AsA; NL: 100 mM NaCl + 0.25 mM LYC; NLA: 100 mM NaCl + 0.25 mM LYC + 0.5 mM AsA.

3.6. Exogenous AsA Regulates the Expression of Genes Related to the SOS Pathway of Tomato Seedlings under Salt Stress

The expression levels of the *SOS1, SOS2, NHX2, NHX3, HKT1;2, VP1*, and *CLC* genes were downregulated to varying degrees in tomato seedling leaves under NaCl treatment compared with control, as shown in Figure 6. The exogenous spraying of AsA significantly upregulated the expression levels of the *SOS1, SOS2, NHX2, NHX3, HKT1;2, VP1*, and *CLC* genes in tomato seedling leaves by 2.12-, 3.35-, 2.64-, 2.19-, 3.05-, 4.55-, 5.55-, 3.76, and 1.44-fold, respectively, compared to the NaCl treatment. The expression of the above genes decreased to different degrees in the NL treatment compared to NaCl stress. However, the NLA treatment reversed this phenomenon, and all the above-mentioned genes were significantly upregulated. This indicates that exogenous AsA can regulate ion homeostasis by regulating the expression of genes related to the SOS pathway.

3.7. Exogenous AsA Regulates Proline (Pro) Content and Its Anabolic Key Enzyme Activities and Gene Expression in Tomato Seedlings under Salt Stress

As shown in Figures 7 and 8, salt stress significantly increased Pro content and the activity and gene expression of P5CS and OAT, while significantly decreasing ProDH activity and gene expression in tomato seedling leaves compared with the control. Compared with the NaCl treatment, the NA treatment significantly reduced Pro content as well as the activities of P5CS and OAT by 60.2%, 45.6, and 14.8%, respectively. The NA treatment downregulated the expression of the P5CS gene by 0.25-fold and the expression of the *OAT* gene by 0.38-fold under NaCl stress. Additionally, the NA treatment significantly increased ProDH activity by 20.7% and upregulated *ProDH* gene expression by 2.64-fold. However, the exogenous application of the AsA inhibitor LYC significantly increased proline content as well as OAT and P5CS activities and their gene expression in tomato seedling leaves under salt stress conditions, while significantly decreasing ProDH activity and gene ex pression. The exogenous spraying of AsA on top of the NL treatment then significantl reversed the trend of the above indicators. These results indicate that exogenous AsA ca reduce the proline content in tomato under salt stress by regulating the activity of ke enzymes and gene expression in proline anabolism.

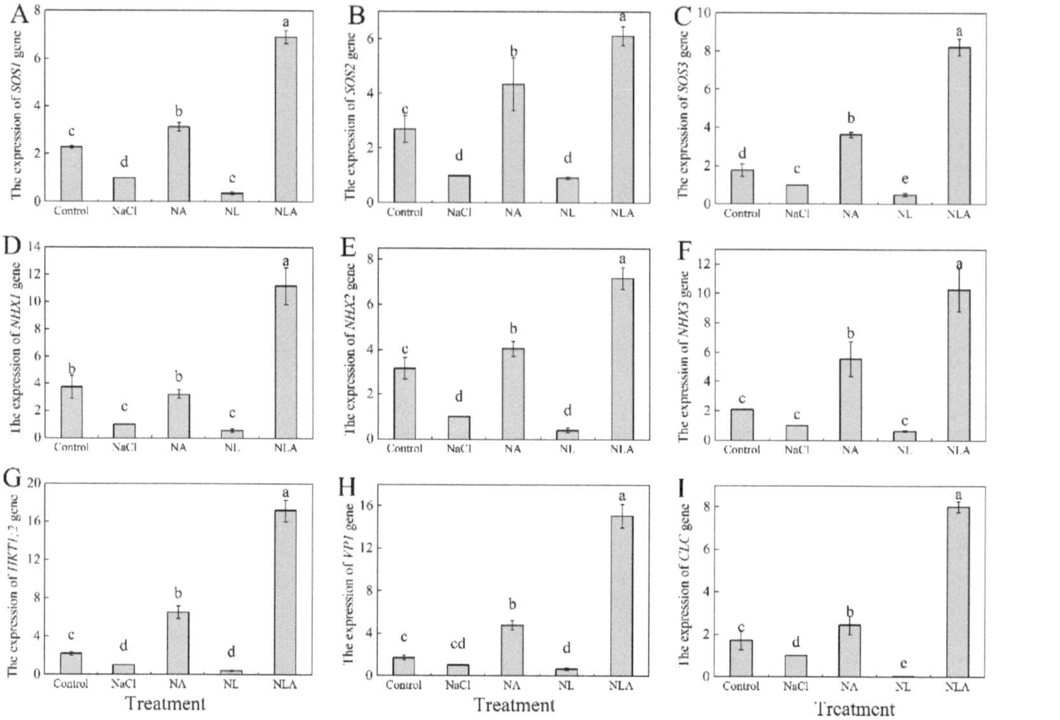

Figure 6. Expression of *SOS1* (Salt overly sensitive 1 gene) (**A**), *SOS2* (Salt overly sensitive 2 gene (**B**), *SOS3* (Salt overly sensitive 3 gene) (**C**), *NHX1* (Na^+/H^+ antiporter 1 gene) (**D**), *NHX2* (Na^+/H^+ antiporter 2 gene) (**E**), *NHX3* (Na^+/H^+ antiporter 3 gene) (**F**), *HKT1;2* (high-affinity potassiun transporter protein gene) (**G**), *VP1* (pyrophosphate-energized vacuolar membrane proton pum gene) (**H**), and *CLC* (chloride channel protein gene) (**I**) genes in the leaves of salt-stressed tomat seedlings with or without exogenous reduced ascorbic acid (AsA) and Lycorine (LYC, AsA synthesi inhibitor) spraying. All measurements were performed on leaves at 3 d after treatment. Error bar represent SD ($n = 3$). Different letters indicate significance differences among treatments ($p < 0.05$ Control: no NaCl and no AsA and no LYC; NaCl: 100 mM NaCl; NA: 100 mM NaCl + 0.5 mM AsA NL: 100 mM NaCl + 0.25 mM LYC; NLA: 100 mM NaCl + 0.25 mM LYC + 0.5 mM AsA.

3.8. Analysis of Correlation

From the results of the correlation analysis, significant positive or negative correlation were observed between some salt-responsive physiological and morphological parameter (Figure 9). For example, aboveground and belowground relative growth rates were sig nificantly and negatively correlated with Cl^- content in plant leaves and roots, and th salt over-sensitive (SOS) regulatory pathways (*SOS1*, *SOS2*, and *SOS3* gene expression were significantly and positively correlated with the expression of the *ProDH* and *GalDH* genes. This suggests that some of the parameters may have similar responses to salt stress However, the significant correlation between two parameters does not imply that they ca be substituted for each other.

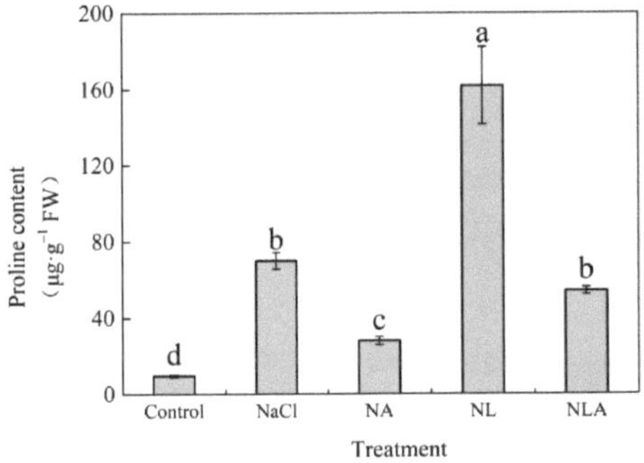

Figure 7. Values of proline content in the leaves of salt-stressed tomato seedlings with or without exogenous reduced ascorbic acid (AsA) and Lycorine (LYC, AsA synthesis inhibitor) spraying. All measurements were performed on leaves at 3 d after treatment. Error bars represent SD (n = 4). Different letters indicate significance differences among treatments ($p < 0.05$). Control: no NaCl and no AsA and no LYC; NaCl: 100 mM NaCl; NA: 100 mM NaCl + 0.5 mM AsA; NL: 100 mM NaCl + 0.25 mM LYC; NLA: 100 mM NaCl + 0.25 mM LYC + 0.5 mM AsA.

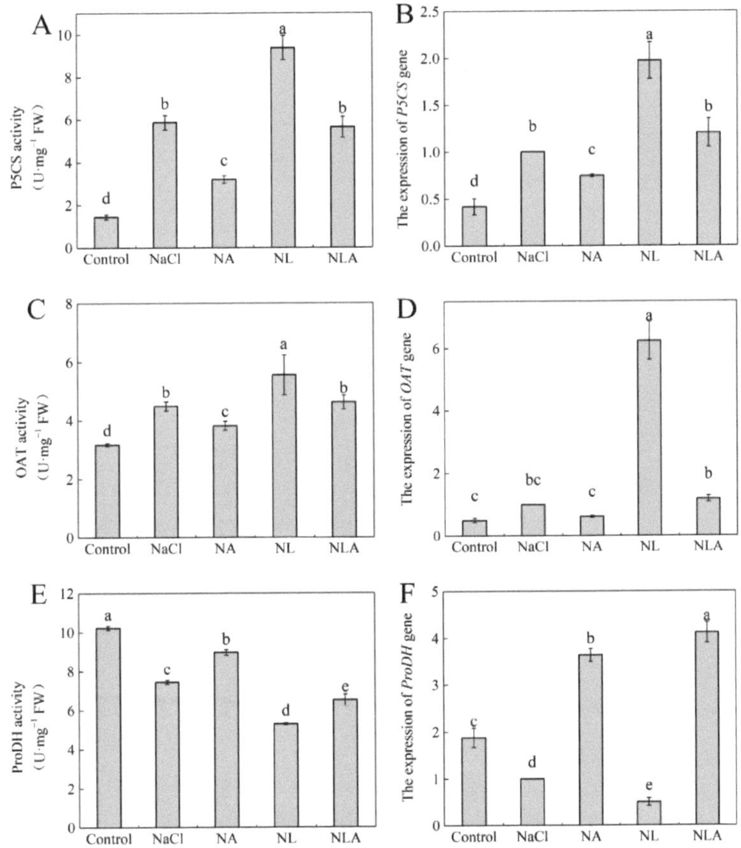

Figure 8. Values of Δ1-pyrroline-5-carboxylate synthase (P5CS) activity (**A**), expression of *P5CS* (Δ1-pyrroline-5-carboxylate synthase gene) gene (**B**), ornithine-δ-aminotransferase (OAT) activity (**C**),

expression of *OAT* (ornithine-δ-aminotransferase gene) gene (**D**), proline dehydrogenase (ProDH) activity (**E**), and expression of *ProDH* (proline dehydrogenase gene) gene (**F**) in the leaves of salt-stressed tomato seedlings with or without exogenous reduced ascorbic acid (AsA) and Lycorine (LYC, AsA synthesis inhibitor) spraying. All measurements were performed on leaves at 3 d after treatment. Error bars represent SD ($n = 4$). Different letters indicate significance differences among treatments ($p < 0.05$). Control: no NaCl and no AsA and no LYC; NaCl: 100 mM NaCl; NA: 100 mM NaCl + 0.5 mM AsA; NL: 100 mM NaCl + 0.25 mM LYC; NLA: 100 mM NaCl + 0.25 mM LYC + 0.5 mM AsA.

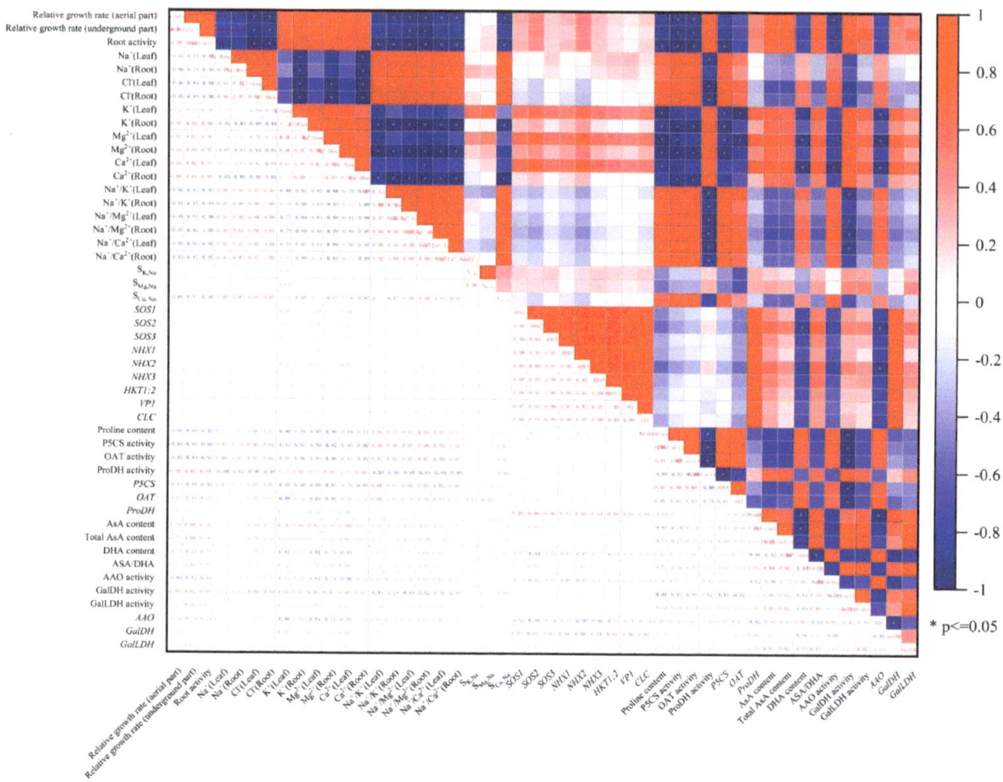

Figure 9. The linear regression coefficients among 48 salt response parameters.

3.9. *Mechanism of AsA in Alleviating Salt Stress*

As shown in Figure 10, salt stress induces a high oxidative state in cells, disrupts ionic homeostasis in the roots and leaves of tomato seedlings, accumulates large amounts of Na^+ and Cl^-, and reduces the expression of genes related to the SOS pathway, thereby disrupting the dynamic balance of proline. These effects ultimately lead to decreased root activity and affect the overall growth and development of the plant. However, the exogenous spraying of AsA significantly reverses these phenomena. Exogenous AsA effectively improves the growth characteristics of plants under salt stress by maintaining high endogenous AsA levels and redox pools. This maintenance mediates the SOS pathway to alleviate ionic toxicity and mitigates osmotic stress by regulating proline synthesis and metabolism. These actions enhance the plant's resilience to stress.

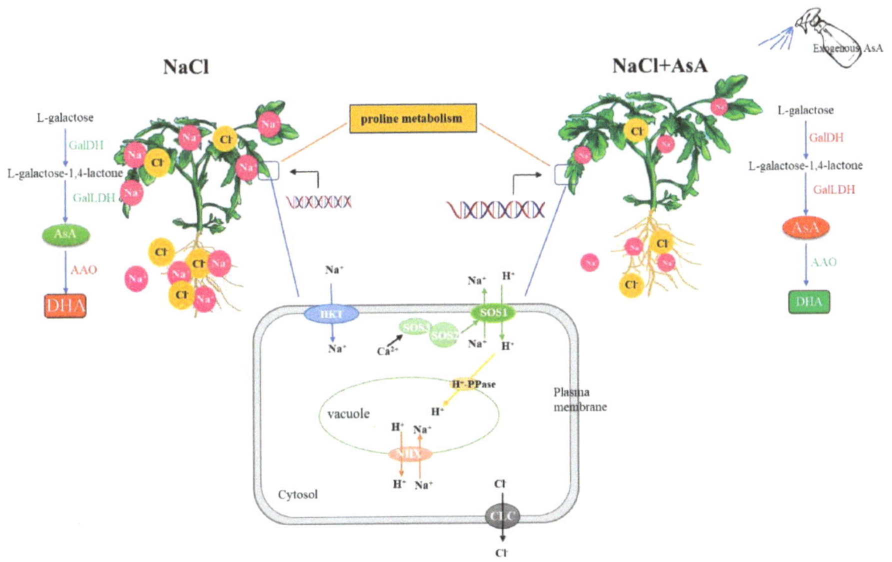

Figure 10. Schematic representation of the mechanism by which exogenous AsA alleviates salt stress in tomato seedlings.

4. Discussion

Salt stress, as a major abiotic stressor, induces ion toxicity, osmotic stress, and oxidative stress, leading to physiological imbalances in plants. These effects severely restrict crop growth, development, and yield, posing a global threat to agricultural production [44]. Plant roots are particularly sensitive to stress signals, and salt stress inhibits their growth, reducing root vitality and impacting the absorption of water and nutrients, thereby affecting the entire plant's normal growth. The findings of this study demonstrate that salt stress significantly inhibited the growth and biomass accumulation of tomato seedlings. However, the exogenous application of AsA effectively improved plant growth characteristics under salt stress by enhancing root vitality, as shown in Figure 1. These results are consistent with earlier reports indicating that AsA can mitigate the adverse effects of salt stress on plant biomass. Furthermore, we observed that exogenous AsA application to tomato seedlings under salt stress increased endogenous AsA levels and the AsA/DHA ratio, as illustrated in Figure 2.

Research has shown that the cellular redox state is an important factor in plants' resistance to abiotic stress [45]. Maintaining a high redox pool of AsA (AsA/DHA ratio) is crucial for plants to scavenge excessive ROS and keep the thiol groups of soluble proteins and membrane proteins in a reduced state. The results indicate that the exogenous application of AsA can effectively promote the growth of tomato seedlings under salt stress, and is positively correlated with endogenous AsA levels (Figure 9). The L-galactose pathway is the main pathway for plants to synthesize AsA, and GalLDH and GalDH are key enzymes in the final two steps of AsA synthesis in the L-galactose pathway. Studies in melon [46] and tobacco [47] have found a close correlation between the content of endogenous AsA and the activity and gene expression of GalLDH; a significant correlation between GalDH activity and AsA content was found in shepherd's purse and Arabidopsis [48]. Studies on corn extracts have found that lycorine (LYC) is an effective inhibitor of GalLDH [49]. To further explore the role of AsA, we applied LYC under salt stress and found that the activities and gene expression of GalLDH and GalDH in tomato seedlings were inhibited, as were the endogenous AsA levels and AsA/DHA ratio (Figures 1 and 3). However, after the exogenous application of AsA, the activities and gene expression of GalDH and GalLDH were significantly increased (Figure 3). AAO is a key enzyme in the AsA oxidation metabolism pathway. When studying the salt tolerance of AAO transgenic plants, it wa found that under normal conditions, there was no obvious phenotype change in AAO ove expressing and antisense transgenic plants compared to the control. However, under hig salt conditions, the germination and photosynthetic rates of the antisense AAO transgeni plants were higher compared to those of the wild type and AAO overexpressing transgeni plants. Furthermore, it was found that the H_2O_2 content in overexpressing transgeni plants was the highest, while the AsA content was very low, indicating that inhibitin the expression of the AAO gene can increase the salt tolerance of plants [50]. This stud found that salt stress led to a significant increase in AAO activity and gene expressio in tomato leaves, while the exogenous spraying of AsA significantly reduced the AA activity and downregulated AAO expression in tomato leaves under NaCl treatment an NL treatment (Figure 3). These findings suggest that the exogenous AsA maintains a hig level of endogenous AsA and redox pool by altering the activities and gene expressio levels of enzymes involved in AsA synthesis and metabolism, thus protecting plants fror oxidative damage caused by salt stress (NaCl and NL treatments) and improving the sa tolerance of tomato seedlings.

Plants maintain ion homeostasis in response to stress environments by reducing exces Na^+ (reducing Na^+ uptake, promoting Na^+ efflux, and compartmentalizing Na^+) to enhanc salt tolerance. Na^+ efflux and compartmentalization are active transport processes mainl driven by the proton gradient generated by H^+-ATPase and H^+-PPase, and mediate by Na^+/H^+ antiporters [51]. Besides Na^+, Cl^- efflux and compartmentalization are als important salt tolerance mechanisms. Studies have shown that exogenous silicon ca reduce Na^+ content in tomato seedling leaves, promote the uptake of K^+, Ca^{2+}, and Mg^{2+} and increase the K^+/Na^+ and Ca^{2+}/Na^+ ratios in leaves by at least 2-fold [52]; exogenou melatonin significantly increases the expression of *SOS* genes (*SOS1*, *SOS2*, and *SOS3*) i plants under salt stress, reducing Na^+ content in the aboveground parts and increasin the K^+/Na^+ ratio [53,54]; exogenous boron, by upregulating the expression of *CLC* gene reduces Cl^- uptake, alleviating the impact of NaCl stress on beet growth [55]. This stud found that salt stress disrupted ion homeostasis in tomato seedling roots and leave resulting in the accumulation of Na^+ and Cl^-, while reducing the uptake of K^+, Ca^{2+}, an Mg^{2+}, affecting the absorption and transport ratios of ions in roots and leaves (Figures and 5), and decreasing the expression of the *SOS* (*SOS1*, *SOS2*, and *SOS3*), *HKT1;2*, *NH* (*NHX1*, *NHX2*, and *NHX3*), *CLC*, and *VP1* genes in seedling leaves (Figure 6). Simila conclusions have been drawn in cabbage [56], cucumber [57], eggplant [58], and othe plants. However, spraying AsA under salt stress and NL treatment revealed that exogenou AsA enhanced Na^+ and Cl^- efflux by upregulating the expression of SOS and CLC gene This treatment also promoted the accumulation of K^+, Ca^{2+}, and Mg^{2+}, resulting in reduce Na^+/K^+, Na^+/Ca^{2+}, and Na^+/Mg^{2+} ratios. Additionally, it reduced Na^+ selectivity i aboveground parts while enhancing the selective absorption and transport capacity of K^+ Ca^{2+}, and Mg^{2+} from roots to leaves (Figures 4 and 5). Regulating *NHX* gene expressio compartmentalized Na^+ into vacuoles, a key mechanism for maintaining water absorptio in plants under salt stress. Regulating *HKT* gene expression recycles Na^+ from transpiratio flow to avoid the excessive accumulation of Na^+ in photosynthetic tissues. Upregulatin *VP1* gene expression provides more driving force for ion transmembrane transport t ensure the normal operation of Na^+/H^+ antiporters, thereby enhancing plant salt toleranc (Figure 6).

Proline, as an osmoprotectant, membrane stabilizer, and ROS scavenger [59–62], an exogenous AsA may have different regulatory patterns in plants under different abioti stresses, with Pro content increasing [63,64] or decreasing [65,66]. Kavi [67] proposed tha maintaining a dynamic balance of Pro under stress conditions is a necessary condition fo normal plant growth and development. Therefore, evaluating the resistance of plants t abiotic stress by Pro content is of great significance. In this study, NaCl stress increase the activity and gene expression of key enzymes in Pro synthesis pathways (P5CS an OAT) in tomato seedling leaves while inhibiting the activity and transcription level of th rate-limiting enzyme in the Pro degradation pathway (ProDH), resulting in a significant increase in Pro content in leaves, indicating that plants have initiated defense mechanisms to resist osmotic stress caused by salt stress (Figures 7 and 8). This is consistent with studies in papaya [68], cucumber [69], and tomato [70]. However, some studies have shown that excess Pro not only fails to alleviate stress-induced damage but also exacerbates growth inhibition under stress [71]. In this study, it was found that the exogenous spraying of the inhibitor LYC dramatically increased Pro content in tomato seedling leaves under salt stress, but the exogenous spraying of AsA reduced excess Pro content in leaves by reducing the synthesis pathway of Pro and enhancing the degradation pathway of Pro, maintaining the dynamic balance of Pro in plants, thereby enhancing the salt tolerance of tomato seedlings.

5. Conclusions

In conclusion, salt stress induces ion imbalance and osmotic stress in tomato seedlings, severely affecting their growth. However, the application of exogenous AsA significantly improves the growth characteristics of plants under salt stress conditions. The research results indicate that exogenous AsA, by regulating the activity and gene expression levels of enzymes related to endogenous AsA synthesis, increases the intracellular AsA content, maintaining the intracellular redox state in a reduced state. Furthermore, exogenous AsA also regulates the transcription levels of multiple ion transporters through the SOS pathway, enhancing the plant's selective absorption of K^+, Ca^{2+}, and Mg^{2+}, thereby alleviating ion imbalance caused by salt stress. Meanwhile, exogenous AsA alleviates osmotic stress by regulating Pro synthesis and metabolism, enhancing the salt tolerance of tomato seedlings (Figure 10).

Author Contributions: The work presented here was carried out in collaboration between all authors. H.L., S.P. and X.C. defined the research theme. X.C. designed methods and experiments, carried out the laboratory experiments, analyzed the data, interpreted the results, and wrote the paper. H.H., Y.C., X.L., W.Z., J.C. and W.X. collaborated on the analysis of ion content and the transcriptional expression assay of key genes of the SOS signaling pathway. H.L. and S.P. conceived and coordinated the study. All authors have read and agreed to the published version of the manuscript.

Funding: This research was funded by the Major Science and Technology Special Projects of Xinjiang Uygur Autonomous Region (No. 2022A02005-2), the Earmarked Fund for XJARS (XJARS-07) and the National Natural Science Foundation of China (Grant No. 31860550).

Data Availability Statement: Data is contained within the article.

Conflicts of Interest: The authors declare no conflicts of interest.

References

1. Alves, R.D.C.; Oliveira, K.R.; Lúcio, J.C.B.; Silva, J.D.S.; Carrega, W.C.; Queiroz, S.F.; Gratão, P.L. Exogenous foliar ascorbic acid applications enhance salt-stress tolerance in peanut plants throughout an increase in the activity of major antioxidant enzymes. *S. Afr. J. Bot.* **2022**, *150*, 759–767. [CrossRef]
2. Chen, X.; Xu, Z.; Zhao, B.; Yang, Y.; Mi, J.; Zhao, Z.; Liu, J. Physiological and Proteomic Analysis Responsive Mechanisms for Salt Stress in Oat. *Front. Plant Sci.* **2022**, *13*, 891674. [CrossRef] [PubMed]
3. Chourasia, K.N.; Lal, M.K.; Tiwari, R.K.; Dev, D.; Kardile, H.B.; Pati, V.U.; Kumar, A.; Vanishree, G.; Kumar, D.; Bhardwaj, V.; et al. Salinity Stress in Potato: Understanding Physiological, Biochemical and Molecular Responses. *Life* **2021**, *11*, 545. [CrossRef] [PubMed]
4. Zelm, E.; Zhang, Y.; Testerink, C. Salt Tolerance Mechanisms of Plants. *Annu. Rev. Plant Biol.* **2020**, *71*, 403–433. [CrossRef] [PubMed]
5. Rahman, A.; Alam, M.U.; Hossain, M.S.; Mahmud, J.A.; Nahar, K.; Fujita, M.; Hasanuzzaman, M. Exogenous gallic acid confers salt tolerance in rice seedlings: Modulation of ion homeostasis, osmoregulation, antioxidant defense, and methylglyoxal detoxification systems. *Agronomy* **2023**, *13*, 16. [CrossRef]
6. Roy, S.J.; Negrão, S.; Tester, M. Salt resistant crop plants. *Curr. Opin. Biotechnol.* **2014**, *26*, 115–124. [CrossRef] [PubMed]
7. Zhao, S.; Zhang, Q.; Liu, M.; Zhou, H.; Ma, C.; Wang, P. Regulation of Plant Responses to Salt Stress. *Int. J. Mol. Sci.* **2021**, *22*, 4609. [CrossRef] [PubMed]
8. Delauney, A.J.; Verma, D.P.S. Proline biosynthesis and osmoregulation in plants. *Plant J.* **1993**, *4*, 215–223. [CrossRef]
9. Mansour, M.M.F.; Ali, E.F. Evaluation of proline functions in saline conditions. *Phytochemistry* **2017**, *140*, 52–68. [CrossRef]

10. Yamada, M.; Morishita, H.; Urano, K.; Shiozaki, N.; Yamaguchi-Shinozaki, K.; Shinozaki, K.; Yoshiba, Y. Effects of free prolin accumulation in petunias under drought stress. *J. Exp. Bot.* **2005**, *56*, 1975–1981. [CrossRef]
11. Li, W.; Sun, J.; Zhang, X.; Ahmad, N.; Hou, L.; Zhao, C.; Pan, J.; Tian, R.; Wang, X.; Zhao, S. The Mechanisms Underlying Sa Resistance Mediated by Exogenous Application of 24-Epibrassinolide in Peanut. *Int. J. Mol. Sci.* **2022**, *23*, 6376. [CrossRe [PubMed]
12. Li, X.; Li, S.; Wang, J.; Lin, J. Exogenous Abscisic Acid Alleviates Harmful Effect of Salt and Alkali Stresses on Wheat Seedling *Int. J. Environ. Res. Public Health* **2020**, *17*, 3770. [CrossRef] [PubMed]
13. Yan, F.; Zhang, J.; Li, W.; Ding, Y.; Zhong, Q.; Xu, X.; Wei, H.; Li, G. Exogenous melatonin alleviates salt stress by improving lea photosynthesis in rice seedlings. *Plant Physiol. Biochem.* **2021**, *163*, 367–375. [CrossRef] [PubMed]
14. Qian, H.F.; Peng, X.F.; Han, X.; Ren, J.; Zhan, K.Y.; Zhu, M. The Stress Factor, Exogenous Ascorbic Acid, Affects Plant Growth an the Antioxidant System in Arabidopsis Thaliana. *Russ. J. Plant Physiol.* **2014**, *61*, 467–475. [CrossRef]
15. Smirnoff, N. Ascorbic Acid Metabolism and Functions: A Comparison of Plants and Mammals. *Free Radic. Biol. Med.* **2018**, *12* 116–129. [CrossRef]
16. Foyer, C.H.; Noctor, G. Ascorbate and Glutathione: The Heart of the Redox Hub. *Plant Physiol.* **2011**, *155*, 2–18. [CrossRe [PubMed]
17. Gallie, D.R. The Role of L-ascorbic acid Recycling in Responding to Environmental Stress and in Promoting Plant Growth. *J. Ex Bot.* **2012**, *64*, 433–443. [CrossRef]
18. Zahra, G.; Marjan, M.; Taher, B.; Ranjbar, M.E. Foliar Application of Ascorbic Acid and Gamma Aminobutyric Acid Can Improv Important Properties of Deficit Irrigated Cucumber Plants (*Cucumis sativus* cv. Us). *Gesunde Pflanz.* **2021**, *73*, 77–84. [CrossRef]
19. Kakan, X.; Yu, Y.; Li, S.; Li, X.; Huang, R.; Wang, J. Ascorbic Acid Modulation by ABI4 Transcriptional Repression of VTC2 in th Salt Tolerance of Arabidopsis. *BMC Plant Biol.* **2021**, *21*, 112. [CrossRef]
20. Crizel, R.L.; Perin, E.C.; Siebeneichler, T.J.; Borowski, J.M.; Messias, R.S.; Rombaldi, C.V.; Galli, V. Abscisic Acid and Stres Induced by Salt: Effect on the Phenylpropanoid, L-Ascorbic Acid and Abscisic Acid Metabolism of Strawberry Fruits. *Plan Physiol. Biochem.* **2020**, *152*, 211–220. [CrossRef]
21. Elkelish, A.; Qari, S.H.; Mazrou, Y.S.A.; Abdelaal, K.A.A.; Hafez, Y.M.; Abu-Elsaoud, A.M.; Batiha, G.E.-S.; El-Esawi, M.A.; E Nahhas, N. Exogenous Ascorbic Acid Induced Chilling Tolerance in Tomato Plants through Modulating Metabolism, Osmolyte Antioxidants, and Transcriptional Regulation of Catalase and Heat Shock Proteins. *Plants* **2020**, *9*, 431. [CrossRef] [PubMed]
22. Hussain, I.; Siddique, A.; Ashraf, M.A.; Rasheed, R.; Ibrahim, M.; Iqbal, M.; Akbar, S.; Imran, M. Does Exogenous Application o Ascorbic Acid Modulate Growth, Photosynthetic Pigments and Oxidative Defense in Okra (*Abelmoschus esculentus* (L.) Moench under Lead Stress? *Acta Physiol. Plant.* **2017**, *39*, 144. [CrossRef]
23. Zhou, X.; Gu, Z.; Xu, H.; Chen, L.; Tao, G.; Yu, Y.; Li, K. The Effects of Exogenous Ascorbic Acid on the Mechanism of Physiologica and Biochemical Responses to Nitrate Uptake in Two Rice Cultivars (*Oryza sativa* L.) under Aluminum Stress. *J. Plant Growt Regul.* **2016**, *35*, 1013–1024. [CrossRef]
24. Shafiq, S.; Akram, N.A.; Ashraf, M.; Arshad, A. Synergistic Effects of Drought and Ascorbic Acid on Growth, Mineral Nutrient and Oxidative Defense System in Canola (*Brassica napus* L.) Plants. *Acta Physiol. Plant.* **2014**, *36*, 1539–1553. [CrossRef]
25. Hura, A.; Ak, B.; Ma, B. Exogenously Applied Ascorbic Acid Alleviates Salt-Induced Oxidative Stress in Wheat. *Environ. Exp. Bo* **2008**, *63*, 224–231. [CrossRef]
26. Shalata, A.; Neumann, P.M. Plants and the Environment. Exogenous Ascorbic Acid (Vitamin C) Increases Resistance to Salt Stres and Reduces Lipid Peroxidation. *J. Exp. Bot.* **2001**, *52*, 2207–2211. [CrossRef] [PubMed]
27. Nagda, J.K.; Bhanu, N.A.; Katiyar, D.; Hemantaranjan, A.; Yadav, D.K. Mitigating Effect of Foliar Applied Ascorbic Acid or Morpho-Physiological, Biochemical Changes and Yield Attributes Induced by Salt Stress in *Vigna radiate*. *Agric. Sci. Dig. Res.* **2017**, *37*, 112–116. [CrossRef]
28. Chen, X.; Zhou, Y.; Cong, Y.; Zhu, P.; Xing, J.; Cui, J.; Xu, W.; Shi, Q.; Diao, M.; Liu, H. Ascorbic Acid-Induced Photosynthetic Adaptability of Processing Tomatoes to Salt Stress Probed by Fast OJIP Fluorescence Rise. *Front. Plant Sci.* **2021**, *12*, 59440 [CrossRef] [PubMed]
29. Van Echelpoel, W.; Boets, P.; Goethals, P.L. Functional Response (FR) and Relative Growth Rate (RGR) Do Not Show the Known Invasiveness of *Lemna minuta* (Kunth). *PLoS ONE* **2016**, *11*, e0166132. [CrossRef]
30. Li, H.S. *Principles and Techniques of Plant Physiological Experiment*; Higher Education Press: Beijing, China, 2000.
31. Nazar, R.; Iqbal, N.; Syeed, S.; Khan, N.A. Salicylic Acid Alleviates Decreases in Photosynthesis under Salt Stress by Enhancing Nitrogen and Sulfur Assimilation and Antioxidant Metabolism Differentially in Two Mungbean Cultivars. *J. Plant Physiol.* **2011** *168*, 807–815. [CrossRef]
32. Epstein, E. How calcium enhances plant salt tolerance. *Science* **1998**, *280*, 1906. [CrossRef] [PubMed]
33. Livak, K.J.; Schmittgen, T.D. Analysis of relative gene expression data using real-time quantitative PCR and the $2^{-\Delta\Delta CT}$ method *Methods* **2001**, *25*, 402–408. [CrossRef] [PubMed]
34. de Freitas, P.A.F.; de Carvalho, H.H.; Costa, J.H.; Miranda, R.D.S.; Saraiva, K.D.D.C.; de Oliveira, F.D.B.; Coelho, D.G.; Prisco, J.T. Gomes-Filho, E. Salt acclimation in sorghum plants by exogenous proline: Physiological and biochemical changes and regulation of proline metabolism. *Plant Cell Rep.* **2019**, *38*, 403–416. [CrossRef] [PubMed]
35. Song, S.Q.; Lei, Y.B.; Tian, X.R. Proline Metabolism and Cross-Tolerance to Salinity and Heat Stress in Germinating Wheat Seeds *Russ. J. Plant Physiol.* **2005**, *52*, 793–800. [CrossRef]

36. Kim, H.R.; Rho, H.W.; Park, J.W.; Park, B.H.; Kim, J.S.; Lee, M.W. Assay of Ornithine Aminotransferase with Ninhydrin. *Anal. Biochem.* **1994**, *223*, 205–207. [CrossRef] [PubMed]
37. Lutts, S.; Kinet, J.M.; Bouharmont, J. NaCl-induced Senescence in Leaves of Rice (*Oryza sativa* L.) Cultivars Differing in Salinity Resistance. *Ann. Bot.* **1996**, *78*, 389–398. [CrossRef]
38. Jiang, M.; Zhang, J. Effect of Abscisic Acid on Active Oxygen Species, Antioxidative Defence System and Oxidative Damage in Leaves of Maize Seedlings. *Plant Cell Physiol.* **2001**, *42*, 1265–1273. [CrossRef] [PubMed]
39. Gatzek, S.; Wheeler, G.L.; Smirnoff, N. Antisense Suppression of L-galactose Dehydrogenase in *Arabidopsis thaliana* Provides Evidence for Its Role in Ascorbate Synthesis and Reveals Light Modulated L-galactose Synthesis. *Plant J.* **2002**, *30*, 541–553. [CrossRef] [PubMed]
40. Ôba, K.; Ishikawa, S.; Nishikawa, M.; Mizuno, H.; Yamamoto, T. Purification and Properties of L-Galactono-γ-Lactone Dehydrogenase, a Key Enzyme for Ascorbic Acid Biosynthesis, from Sweet Potato Roots. *J. Biochem.* **1995**, *117*, 120–124. [CrossRef]
41. Esaka, M.; Hattori, T.; Fujisawa, K.; Sakajo, S.; Asahi, T. Molecular Cloning and Nucleotide Sequence of fullFull-length cDNA for Ascorbate Oxidase from Cultured Pumpkin Cells. *Eur. J. Biochem.* **1990**, *191*, 537–541. [CrossRef]
42. Li, X.; Wang, S.; Chen, X.; Cong, Y.; Cui, J.; Shi, Q.; Liu, H.; Diao, M. The positive effects of exogenous sodium nitroprusside on the plant growth, photosystem II efficiency and Calvin cycle of tomato seedlings under salt stress. *Sci. Hortic.* **2022**, *299*, 111016. [CrossRef]
43. Zhang, W.; He, X.; Chen, X.; Han, H.; Shen, B.; Diao, M.; Liu, H.Y. Exogenous selenium promotes the growth of salt-stressed tomato seedlings by regulating ionic homeostasis, activation energy allocation and CO_2 assimilation. *Front. Plant Sci.* **2023**, *14*, 1206246. [CrossRef] [PubMed]
44. Ntanasi, T.; Karavidas, I.; Zioviris, G.; Ziogas, I.; Karaolani, M.; Fortis, D.; Conesa, M.À.; Schubert, A.; Savvas, D.; Ntatsi, G. Assessment of Growth, Yield, and Nutrient Uptake of Mediterranean Tomato Landraces in Response to Salinity Stress. *Plants* **2023**, *12*, 3551. [CrossRef] [PubMed]
45. Dietz, K.J. Redox-dependent regulation, redox control and oxidative damage in plant cells subjected to abiotic stress. *Methods Mol. Biol.* **2010**, *639*, 57–70. [CrossRef] [PubMed]
46. Pateraki, I.; Sanmartin, M.; Kalamaki, M.S.; Gerasopoulos, D.; Kanellis, A.K. Molecular Characterization and Expression Studies during Melon Fruit Development and Ripening of L-galactono-1, 4-lactone Dehydrogenase. *J. Exp. Bot.* **2004**, *55*, 1623–1633. [CrossRef] [PubMed]
47. Tabata, K.; Takaoka, T.; Esaka, M. Gene Expression of Ascorbic Acid-Related Enzymes in Tobacco. *Phytochemistry* **2002**, *61*, 631–635. [CrossRef] [PubMed]
48. Zheng, Q.; Yang, Z.; Zhang, X.; Zhang, F.; Hao, L.Z. Effects of drought stress on ascorbic acid contents and metabolism related enzymes of *Pugionium cornutum* and *P. dolabratum*. *Plant Physiol. J.* **2018**, *54*, 1865–1874. [CrossRef]
49. Arrigoni, O.; Gara, L.D.; Paciolla, C.; Evidente, A.; de Pinto, M.C.; Liso, R. Lycorine: A powerful inhibitor of L-galactono-γ-lactone dehydrogenase activity. *J. Plant Physiol.* **1997**, *150*, 362–364. [CrossRef]
50. Yamamoto, A.; Bhuiyan, M.N.; Waditee, R.; Tanaka, Y.; Esaka, M.; Oba, K.; Jagendorf, A.; Takabe, T. Suppressed expression of the apoplastic ascorbate oxidase gene increases salt tolerance in tobacco and Arabidopsis plants. *J. Exp. Bot.* **2005**, *56*, 1785–1796. [CrossRef]
51. Toranj, S.; Aliabad, K.K.; Abbaspour, H.; Saeedpour, A. Effect of Salt Stress on the Genes Expression of the Vacuolar H^+-Pyrophosphatase and Na^+/H^+ Antiporter in *Rubia tinctorum*. *Mol. Biol. Rep.* **2020**, *47*, 235–245. [CrossRef]
52. Yunus, Q.; Zari, M. Effect of Exogenous Silicon on Ion Distribution of Tomato Plants under Salt Stress. *Commun. Soil Sci. Plant Anal.* **2017**, *48*, 1843–1851. [CrossRef]
53. Ren, J.; Ye, J.; Yin, L.; Li, G.; Deng, X.; Wang, S. Exogenous Melatonin Improves Salt Tolerance by Mitigating Osmotic, Ion, and Oxidative Stresses in Maize Seedlings. *Agronomy* **2020**, *10*, 663. [CrossRef]
54. Zhang, T.; Shi, Z.; Zhang, X.; Zheng, S.; Wang, J.; Mo, J. Alleviating Effects of Exogenous Melatonin on Salt Stress in Cucumber. *Sci. Hortic.* **2020**, *262*, 109070. [CrossRef]
55. Dong, X.; Sun, L.; Guo, J.; Liu, L.; Han, G.; Wang, B. Exogenous Boron Alleviates Growth Inhibition by NaCl Stress by Reducing Cl-Uptake in Sugar Beet (*Beta vulgaris*). *Plant Soil* **2021**, *464*, 423–439. [CrossRef]
56. Wang, J.; Qiu, N.; Wang, P.; Zhang, W.; Yang, X.; Chen, M.; Wang, B.; Sun, J. Na^+ Compartmentation Strategy of Chinese Cabbage in Response to Salt Stress. *Plant Physiol. Biochem.* **2019**, *140*, 151–157. [CrossRef] [PubMed]
57. Yu, J.; Yu, J.; Liao, W.; Xie, J.; Niu, L.; Zhang, G.; Lv, J.; Xiao, X.; Wu, Y. Ethylene was involved in Ca^{2+}-regulated Na^+ homeostasis, Na^+ transport and cell ultrastructure during adventitious rooting in cucumber explants under salt stress. *J. Plant Biol.* **2020**, *63*, 311–320. [CrossRef]
58. Brenes, M.; Solana, A.; Boscaiu, M.; Fita, A.; Vicente, O.; Calatayud, Á.; Prohens, J.; Plazas, M. Physiological and Biochemical Responses to Salt Stress in Cultivated Eggplant (*Solanum melongena* L.) and in *S. insanum* L. a Close Wild Relative. *Agronomy* **2020**, *10*, 651. [CrossRef]
59. Ashraf, M.; Foolad, M.R. Roles of Glycine Betaine and Proline in Improving Plant Abiotic Stress Resistance. *Environ. Exp. Bot.* **2007**, *59*, 206–216. [CrossRef]
60. Ghoulam, C.; Foursy, A.; Fares, K. Effects of Salt Stress on Growth, Inorganic Ions and Proline Accumulation in Relation to Osmotic Adjustment in Five Sugar Beet Cultivars. *Environ. Exp. Bot.* **2002**, *47*, 39–50. [CrossRef]

61. Hinai, M.S.A.; Ullah, A.; Al-Rajhi, R.S.; Farooq, M. Proline Accumulation, Ion Homeostasis and Antioxidant Defence Syste Alleviate Salt Stress and Protect Carbon Assimilation in Bread Wheat Genotypes of Omani Origin. *Environ. Exp. Bot.* **2022**, *19* 104687. [CrossRef]
62. Mattioli, R.; Palombi, N.; Funck, D.; Trovato, M. Proline Accumulation in Pollen Grains as Potential Target for Improved Yiel Stability Under Salt Stress. *Front. Plant Sci.* **2020**, *11*, 582877. [CrossRef] [PubMed]
63. El-Bassiouny HM, S.; Sadak, M.S. Impact of Foliar Application of Ascorbic Acid and α-Tocopherol on Antioxidant Activity an Some Biochemical Aspects of Flax Cultivars under Salinity Stress. *Acta Biológica Colomb.* **2014**, *20*, 209–222. [CrossRef]
64. Farooq, M.; Irfan, M.; Aziz, T.; Ahmad, I.; Cheema, S.A. Seed Priming with Ascorbic Acid Improves Drought Resistance of Whe *J. Agron. Crop Sci.* **2013**, *199*, 12–22. [CrossRef]
65. Darvishan, M.; Tohidi Moghadam, H.R.; Zahedi, H. The Effects of Foliar Application of Ascorbic Acid (Vitamin C) on Physiologic and Biochemical Changes of Corn (*Zea mays* L.) under Irrigation Withholding in Different Growth Stages. *Maydica* **2013**, *5* 195–200.
66. Terzi, R.; Kalaycıoglu, E.; Demiralay, M.; Saglam, A.; Kadioglu, A. Exogenous Ascorbic Acid Mitigates Accumulation of Abscis Acid, Proline and Polyamine under Osmotic Stress in Maize Leaves. *Acta Physiol. Plant.* **2015**, *37*, 43. [CrossRef]
67. Kavi Kishor, P.B.; Sreenivasulu, N. Is Proline Accumulation Per Se Correlated with Stress Tolerance or is Proline Homeostasis More Critical Issue? *Plant Cell Environ.* **2014**, *37*, 300–311. [CrossRef] [PubMed]
68. Álvarez-Méndez, S.J.; Urbano-Gálvez, A.; Mahouachi, J. Mitigation of salt stress damages in *Carica papaya* L. seedlings throug exogenous pretreatments of gibberellic acid and proline. *Chil. J. Agric. Res.* **2022**, *82*, 167–176. [CrossRef]
69. Naliwajski, M.; Skłodowska, M. The Relationship between the Antioxidant System and Proline Metabolism in the Leaves c Cucumber Plants Acclimated to Salt Stress. *Cells* **2021**, *10*, 609. [CrossRef]
70. Shin, Y.K.; Bhandari, S.R.; Jo, J.S.; Song, J.W.; Cho, M.C.; Yang, E.Y.; Lee, J.G. Response to Salt Stress in Lettuce: Changes i Chlorophyll Fluorescence Parameters, Phytochemical Contents, and Antioxidant Activities. *Agronomy* **2020**, *10*, 1627. [CrossRe
71. Suekawa, M.; Fujikawa, Y.; Esaka, M. Exogenous Proline Has Favorable Effects on Growth and Browning Suppression in Rice bu Not in Tobacco. *Plant Physiol. Biochem.* **2019**, *142*, 1–7. [CrossRef]

plants

Article

Testing a Simulation Model for the Response of Tomato Fruit Quality Formation to Temperature and Light in Solar Greenhouses

Yongdong Qin [1,†], Ao Gong [2,†], Xigang Liu [1], Nan Li [1], Tuo Ji [1,3], Jing Li [1,3,4,*] and Fengjuan Yang [1,3,4,*]

1 College of Horticulture Science and Engineering, Shandong Agricultural University, Tai'an 271018, China
2 College of Information Science and Engineering, Shandong Agricultural University, Tai'an 271018, China
3 Key Laboratory of Biology and Genetic Improvement of Horticultural Crop (Huang-Huai Region), Ministry of Agriculture and Rural Affairs, Tai'an 271018, China
4 Shandong Collaborative Innovation Center for Fruit and Vegetable Production with High Quality and Efficiency, Tai'an 271018, China
* Correspondence: lij19900525@163.com (J.L.); fjyang@sdau.edu.cn (F.Y.)
† These authors contributed equally to this work.

Abstract: Temperature and light are the key factors affecting the formation of tomato fruit quality in greenhouse cultivation. However, there are few simulation models that examine the relationship between tomato fruit quality formation and temperature and light. In this study, a model was established that investigated the relationships between soluble sugar (SSC), organic acid content (OAC), and SSC/OAC and the cumulative product of thermal effectiveness and photosynthetically active radiation (TEP) during the fruit-ripening period in a solar greenhouse. The root mean square error (RMSE) values were calculated to compare the consistency between the simulated and measured values, and the RMSE values for SSC, OAC, and SSC/OAC were 0.09%, 0.14%, and 0.358, respectively. The combined weights of quality indicators were obtained using the analytic hierarchy process (AHP) and entropy weighting method, ranking as SSC > OAC > SSC/OAC > CI > lycopene > Vc > fruit firmness. The comprehensive fruit quality evaluation value was obtained using the TOPSIS method (Technique for Order Preference by Similarity to an Ideal Solution) and a simulation model between comprehensive tomato fruit quality and TEP was explored. This study could accurately simulate and quantify the accumulation of tomato fruit quality during fruit ripening in response to environmental conditions in a solar greenhouse.

Keywords: tomato; fruit quality; light and temperature; simulation model; TEP

check for updates

Citation: Qin, Y.; Gong, A.; Liu, X.; Li, N.; Ji, T.; Li, J.; Yang, F. Testing a Simulation Model for the Response of Tomato Fruit Quality Formation to Temperature and Light in Solar Greenhouses. *Plants* **2024**, *13*, 1662. https://doi.org/10.3390/plants13121662

Academic Editors: Weijie Jiang and Wenna Zhang

Received: 25 April 2024
Revised: 6 June 2024
Accepted: 13 June 2024
Published: 15 June 2024

1. Introduction

Tomatoes are one of the most grown vegetables in the world, with a global annual output of 170 million t. As one of the main production areas, China's tomato-planting area was 1.109 million hm^2 in 2018, with a yield of 64.832 million t [1,2]. With the abundant supply of vegetables, consumers increasingly demand high-quality vegetables. Tomatoes have become one of the most popular vegetables because of their rich minerals, vitamins, organic acids, essential amino acids, and other nutrients [3,4].

Environmental factors in the greenhouse are the key factors affecting the growth and development of tomatoes. In recent years, with the global warming, extreme weather events frequently occur, which have a serious impact on the growth, development, yield, and quality of crops [5,6]. Temperature is one of the key environmental factors affecting crop yield and quality; different crops have their optimum temperature ranges. High temperatures cause accelerated ripening of tomatoes, so average harvested fruit weights are often reduced [7,8]. Yield and quality are also affected by low temperatures and the fruit-ripening time is prolonged when plants are grown at lower temperatures [9,10]. Changes in daily temperature patterns may affect the growth and metabolism of plants and fruit, and ultimately affect fruit quality [11,12]. At high temperatures (30–35 °C), SSC increases, while OAC decreases [13]. The biosynthesis of lycopene is strongly inhibited at temperatures below 12 °C, while temperatures above 32 °C hinder this process and lead to a decrease in its content [14,15], reaching a maximum concentration at 25 °C [16]. Studies have shown that the vitamin C concentration is seen to increase when plants are high-temperature stressed at the flowering stage [17]. Light is the driving force of plant photosynthesis, and the accumulation of photosynthetic assimilates is crucial to yield [18]. In addition, light can affect plant transpiration by regulating the closure of leaf stomata, thus influencing crop transport and growth [19,20]. A prolonged light treatment increases the starch and sugar contents in tomato fruit, and promotes the synthesis of anthocyanins in apple fruit skin and the accumulation of soluble sugars in apple flesh [21,22]. Extending the light duration not only promotes fruit coloration, but also the accumulation of fruit sugar; the appearance quality and fresh eating quality of fruit are significantly improved [23,24]. In the life activities of plants, there is a complex interaction between temperature signals and light signals, and temperature and light together affect the growth and development of plants [25–27].

Greenhouse cultivation is one of the main methods for tomato production, but it often faces problems such as high temperatures in summer and low temperatures and light in winter, which seriously affects tomato yield and quality. The interaction of multiple environmental factors affects the growth and development of tomatoes. With accumulating solar radiation (CSR), cumulative heat unit (CHU), and steam pressure deficiency (VPD) as variables, a multi-linear regression (MLR) method was used to establish a model for fertility, flowers, and fruit [28]. Wu et al. (2021) established a relationship model of above-ground biomass based on GDDs (growing degree days) and then analyzed the influence of three irrigation regimes on greenhouse-grown tomatoes [29]. Correlations between tomato seedling quality characteristics (root–shoot ratio, G value, and healthy indices) and TEP (thermal effectiveness and photosynthetically active radiation) have been explored to establish models [30]. Most previous studies have mainly focused on exploring the optimal combination of temperature and light environments at the tomato seedling stage exploring the relationship between tomato growth and development and environmental factors [31,32]. However, fruit quality is the main source of value of tomatoes and there are few reports on the demand characteristics of environmental factors for fruit quality during the fruit-ripening period.

Tomato fruit quality is a comprehensive concept that is the result of the joint action of various quality indicators [33]. It mainly includes the appearance quality, taste quality, and nutritional quality [34]. Soluble sugar and acid jointly determine the taste quality of tomatoes [35]. Lycopene and Vc are the main nutrients in tomato fruit and have many health benefits. Lycopene represents 80 percent of the carotenoids in tomatoes and Vc has reducing and chelating properties, helping to enhance the body's absorption of iron [36,37]. The appearance quality of tomato fruit (color, size, etc.) is also an important factor when determining the quality of tomato products. Many methods have been used to evaluate tomato quality such as the analytic hierarchy process (AHP), principal component analysis (PCA), Gray relational analysis (GAR), and entropy method as well as TOPSIS (Technique for Order Preference by Similarity to Ideal Solution) [38,39]. However, some methods are influenced by subjective factors and others are evaluated based on raw data resulting in inaccurate results. Therefore, a combination of multiple evaluation methods can obtain more accurate evaluation results. Based on GAR and PCA, a comprehensive fruit quality grade was calculated using the combined evaluation method [34]. By combining the weights of AHP and the entropy method to obtain combined weights and using a multi-level fuzzy evaluation to obtain comprehensive evaluation indicators, a multi-factor regulation model of water and fertilizer for the comprehensive growth of cherry tomatoes was constructed [40]. The combined weights were obtained by combining AHP and the entropy method weights using game theory and TOPSIS was used to comprehensively evaluate different experimental treatments. Thus, a multi-factor coupled regulation model was constructed for a greenhouse environment based on the comprehensive growth of cherry-tomato seedlings [32].

In this study, we planted 10 batches of tomatoes throughout the first and second half of the year in a solar greenhouse with natural environmental conditions. The relationship between the appearance quality, taste quality, and nutritional quality of tomato fruit and temperature and light conditions was analyzed. A suitable range of temperature and light conditions for tomato ripening was determined. A model was established that examined the relationships between soluble sugar (SSC), organic acid content (OAC), and SSC/OAC in tomato fruit and the cumulative product of TEP during the fruit-ripening period in a solar greenhouse. A comprehensive evaluation analysis was also performed. Our results uncovered the demand characteristics of temperature and light environments during the fruit-ripening period, providing a theoretical basis for the production of high-quality tomatoes in a solar greenhouse.

2. Materials and Methods

2.1. Plant Materials and Growth Conditions

Tomato cultivation was conducted in the solar greenhouse of the Science and Technology Innovation Park of Shandong Agricultural University. 'Kaideyali 1832', one of the commonly used cultivated tomato varieties in Shandong Province, China, was selected as the experimental material. The planting dates are shown in Table 1. A Yamasaki tomato nutrient solution was selected for fertilizer and water management using a regular drip irrigation of water and fertilizer integrated machine [41].

Table 1. Dates of tomato planting.

	Planting Dates		Planting Date
T1	18 March 2021	T6	13 August 2021
T2	30 March 2021	T7	28 August 2021
T3	13 April 2021	T8	8 September 2021
T4	25 April 2021	T9	19 September 2021
T5	8 May 2021	T10	30 September 2021

2.2. Measurements

2.2.1. Meteorological Data

Environmental data such as temperature and solar radiation in the solar greenhouse were automatically collected by "Shennong IOT" equipment, developed by the Big Data Center of Shandong Agricultural University. In total, 6 sensors were evenly distributed in the solar greenhouse. Their positions were set at the tomato apexes and the middle of the canopy, and the data were uploaded every 5 min.

Temperature and solar radiation are the two most important factors for tomato fruit development. To consider the combined effect of the two factors, we employed the concept of an accumulated production of photosynthetically active radiation and relative thermal effectiveness (TEP). TEP can be calculated as follows [42].

$$\mathrm{RTE} = \begin{cases} 0 & (\mathrm{T} < \mathrm{T_b}) \\ (\mathrm{T} - \mathrm{T_b})/(\mathrm{T_{ab}} - \mathrm{T_b}) & (\mathrm{T_b} \leq \mathrm{T} < \mathrm{T_{ab}}) \\ 1 & (\mathrm{T_{ab}} \leq \mathrm{T} \leq \mathrm{T_{ou}}) \\ (\mathrm{T_m} - \mathrm{T})/(\mathrm{T_m} - \mathrm{T_{ou}}) & (\mathrm{T_{ou}} < \mathrm{T} \leq \mathrm{T_m}) \\ 0 & (\mathrm{T} < \mathrm{T_m}) \end{cases} \quad (1)$$

$$\mathrm{HTEP} = \mathrm{RTE} \times \mathrm{PAR} \times 3600 \times 10^{-6} \quad (2)$$

$$\mathrm{DTEP} = \sum(\mathrm{HTEP}) \quad (3)$$

$$\mathrm{TEP}_{(i+1)} = \mathrm{TEP}_{(i)} + \mathrm{DTEP}_{(i)} \quad (4)$$

In Formula (1), T is the average temperature per hour (°C), T_b is the lower critical growth temperature (°C), T_{ob} is the lower critical optimum temperature (°C), T_{ou} is the higher critical optimum temperature (°C), and T_m is the higher critical temperature (°C). During the fruit-ripening period, T_b = 15 °C, T_{ob} = 22 °C, T_{ou} = 28 °C, and T_m = 35 °C.

In Formula (2), HTEP is the hourly production of thermal effectiveness and PAR in MJ/(m^2 h).

In Formula (3), DTEP is the daily production of thermal effectiveness and PAR in MJ/(m^2 d).

In Formula (4), $TEP_{(i)}$ is TEP after day i in MJ/m^2 and $TEP_{(i+1)}$ is TEP after i + 1 day in MJ/m^2.

2.2.2. Fruit Quality Parameters

In the second half of the year, tomatoes were collected at 5-day intervals from the time of color change until they were fully ripe (red) and 4 tomatoes of a uniform size and free from external pests and diseases were randomly selected at each time for the determination of fruit quality indicators. First, the color index was measured and then the fruit firmness was measured. Finally, the fruit was crushed with a sampler, quickly frozen with liquid nitrogen, and stored in an ultra-low-temperature refrigerator (−80 °C) to determine the biochemical indicators.

Appearance Quality Parameters

A digital fruit hardness tester (STEP Systems GmbH, Nuremberg, Germany) was used to measure the hardness of tomatoes. The color difference value was measured using a color difference meter (NR110; Shenzhen Tianyouli Standard Light Source Co., Ltd. Shenzhen, China). Three measurement points (evenly distributed) were selected in the tomato equatorial direction and the average value of the three measurement points was used as the CIE color space index of tomato fruit. L* represented black and white, a* represented red and green, and b* represented yellow and blue. It was then converted into a fruit color index (CI) [34].

$$CI = 2000 \times a/(L^*(a^{*2} + b^{*2})^{0.5}) \quad (5)$$

Taste Quality Parameters

The soluble sugar content (SSC) was measured using the anthrone–sulfuric acid colorimetric method. Organic acid content (OAC) was titrated using a 0.1 M NaOH solution [43].

Nutrient Quality Parameters

Lycopene was measured using the spectrophotometric method. The vitamin C content (Vc) was measured using the 2, 6-dichloroindophenol titration method [43].

2.3. *Comprehensive Evaluation Method of Tomato Fruit Quality*

A hierarchical model for a comprehensive evaluation model of tomato fruit quality was established (Figure 1). All fruit quality indicators were divided into appearance quality, taste quality, and nutritional quality, which constituted the criterion layer. All secondary indicators were classified and recorded as sub-factors, which together constituted the scheme layer, including the fruit color index, firmness; SSC, OAC, SSC/OAC, lycopene and Vc.

2.3.1. Determination of Factor (Subjective) Weights via AHP

The analytic hierarchy process (AHP) is a technique and method that combines quantitative and qualitative methods to calculate decision weights to solve complex multi-factor problems. The method establishes a judgment matrix by measuring the relative importance of multiple factors, transforms the problem into the relative importance of comparative indicators, and determines the weight of each factor and sub-factor of the hierarchical model by quantitatively comparing each indicator. The indicator values in the judgment matrix are determined using a scale from 1 to 9 [40,44].

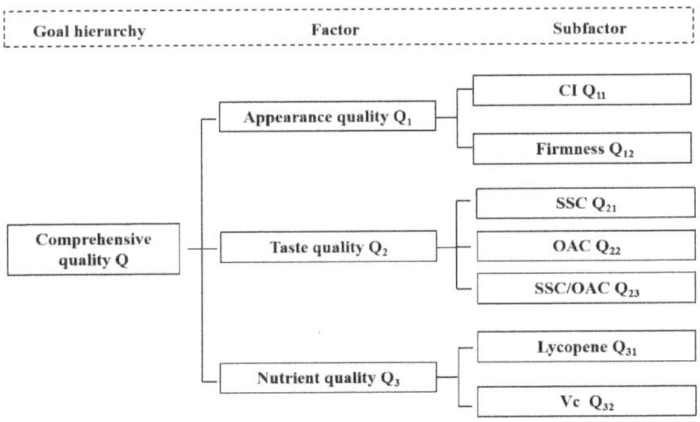

Figure 1. Evaluation hierarchy of comprehensive evaluation of tomato quality.

2.3.2. Determination of Sub-Factor (Objective) Weights Using the Entropy Method

The entropy method is often used to determine the objective weight of an index. The greater the amount of information contained in the data, the smaller the entropy, indicating that the index has a greater impact on a comprehensive evaluation the higher the weight [36]. For the calculation of sub-factor weights using the entropy method, the measured data of the sub-factor set were initially standardized as follows:

$$r_{ij} = (X_{ij} - \min\{X_{ij}\}) / (\max\{X_{ij}\} - \min\{X_{ij}\}) \ (i = 1,2,3,\ldots,n;\ j = 1,2,3,\ldots,m) \quad (6)$$

where r_{ij} donates the standardized value. Subsequently, the specific gravity of these indices could be calculated as follows:

$$P_{ij} = r_{ij} / \sum_{i=1}^{n} r_{ij} \quad (7)$$

The i-th factor and j sub-factor information entropy (e_j) were defined as:

$$e_j = -\frac{1}{\ln n} \sum_{i=1}^{n} p_{ij} \ln p_{ij} \quad (8)$$

Finally, the weight (w_{ij}) of the j sub-factor was determined as follows:

$$w_{ij} = (1 - e_j) / \left(\sum_{j=1}^{m} (1 - e_j) \right) \quad (9)$$

2.3.3. Calculation of Combined Weights

Combined weights were obtained by multiplying the factor (subjective) weights obtained from AHP and the sub-factor (objective) weights obtained from the entropy method.

2.3.4. Comprehensive Evaluation Based on TOPSIS

The TOPSIS method is a commonly used comprehensive evaluation method that can make full use of original data information to accurately reflect the advantages and disadvantages of each scheme.

After data standardization, a weighted decision matrix (Z) based on combination weights was established as follows:

$$z = r_{ij} w_{ij} \quad (10)$$

where r_{ij} represents the measured data after standardization and w_{ij} represents the comprehensive weights based on game theory.

Then, the optimal and worst vectors constituted by the maximum and minimum values of each column of the weighted decision matrix were denoted as:

$$Z^{+} = (Z_{max1}, Z_{max2}, \cdots Z_{maxn}) \quad (11)$$

$$Z^{-} = (Z_{min1}, Z_{min2}, \cdots Z_{minn}) \quad (12)$$

where Z is the weighted decision matrix.

The distance between the weighted decision matrices Z^+ and Z^- (D^+ and D^-) was calculated and, finally, the similarity of the i-th factor (C_j) was calculated.

$$D_i^{+} = \sqrt{\sum_{j=1}^{m}\left(Z - Z_j^{+}\right)^2} \quad (13)$$

$$D_i^{-} = \sqrt{\sum_{j=1}^{m}\left(Z - Z_j^{-}\right)^2} \quad (14)$$

$$C_i = D_i^{-} / (D_i^{+} + D_i^{-}) \quad (15)$$

2.4. Evaluation of Simulated Performance

In this study, the root mean square error (RMSE) was used to evaluate the reliability of the model. RMSE represents the relative error and absolute error between measured and simulated values, respectively [45]. The smaller the RMSE value, the higher the consistency between the simulated value and the measured value, indicating that a model can accurately and reliably predict results.

$$RMSE = \sqrt{\frac{\sum_{i=1}^{n}(SIM_i - OBS_i)^2}{n}} \quad (16)$$

where OBS_i represents the measured data, SIM_i represents the predicted data, and n refers to the number of samples.

2.5. Statistical Analysis

The experimental data were processed using Excel 2010. The weight of the entropy method was calculated using Python. The weight of AHP and the TOPSIS analysis were calculated using DPS.

3. Results

3.1. Environmental Data and Growth Characteristics of Tomato Quality

3.1.1. Variations in Environmental Factors in the Solar Greenhouse throughout the Year

In order to understand the variations in temperature and light conditions in the solar greenhouse, we monitored environmental changes in real time from March 2021 to February 2022. As shown in Figure 2a, the daily mean air temperature increased from 17.62 °C in March 2021 to 33.99 °C in July and August 2021, then decreased to 10.75 °C in January 2022. The average temperature was above 18 °C in July and August, and the night temperature was above 15 °C from May to September. The durations of the daytime temperatures > 30 °C and night-time temperatures > 22 °C gradually increased, reaching a maximum in July before gradually decreasing. The longest durations of the most suitable daytime temperatures (18–30 °C) and night-time temperatures (15–22 °C) for tomato development were in April, September, and October, while the shortest were in July (Table S1).

As shown in Figure 2b, there were daily fluctuations in solar radiation throughout the tomato cultivation period. Significant solar radiation was recorded in the first half of the year, fluctuating around 9 $MJ \cdot m^2 \cdot d^{-1}$; in the second half of the year, solar radiation decreased from August to the end of October, then remained stable (approximately 8 $MJ \cdot m^2 \cdot d^{-1}$). This may have been caused by the shorter duration of daylight in winter and the covering of insulation at night. The average monthly sunshine hours increased and then decreased throughout the year, with June and July having the longest sunshine hours at 13.00 h and 13.45 h, respectively. January had the shortest average daylight hours at 7.65 h (Table S2).

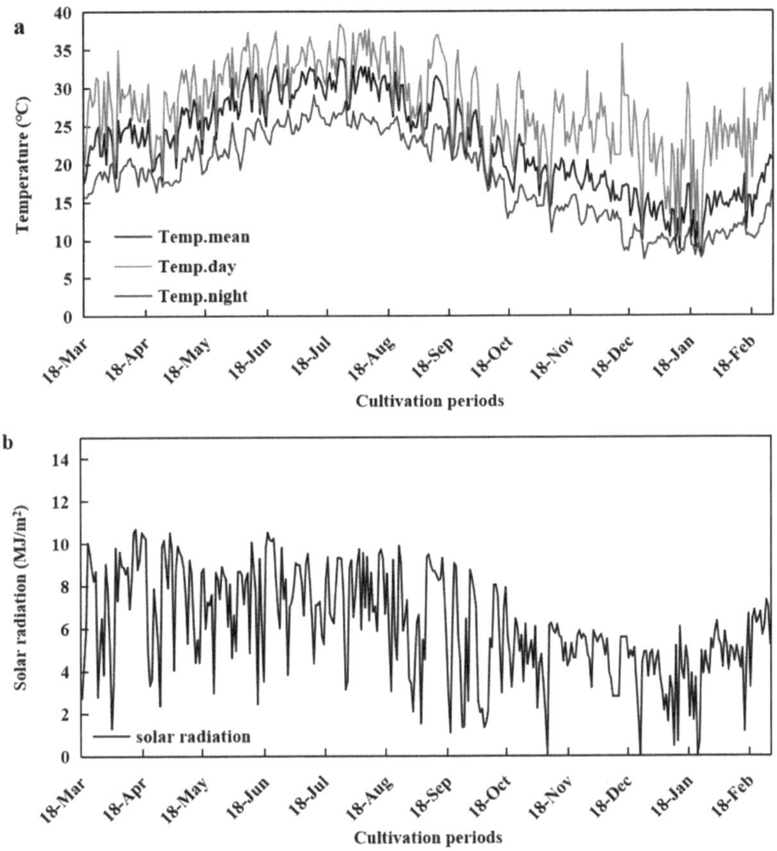

Figure 2. Variations in temperature (**a**) and solar radiation (**b**) in a solar greenhouse throughout the year. Temp.mean represents the average daily temperature, Temp.day represents the average daytime temperature, and Temp.night represents the average night-time temperature.

3.1.2. Tomato Fruit Quality at Different Planting Periods

In order to explore suitable temperature and light environment ranges for fruit quality formation during the ripening period, we planted 10 batches of tomatoes throughout the first and second half of the year in a solar greenhouse with natural environmental conditions (Table 1). The results showed that there was no significant difference in the color indices and SSC of tomatoes from different planting batches, while significant differences were found in OAC, Vc, lycopene, and fruit firmness (Table 2). As a single index cannot reflect the comprehensive quality of tomatoes, we conducted a comprehensive evaluation and analysis. Based on AHP and the entropy weight method, the comprehensive weights of the fruit quality indicators were obtained (Table 3). The TOPSIS method was used to obtain the comprehensive evaluation value of the tomatoes. The results showed that T6 had the highest comprehensive quality and the appropriate temperature and light ranges for the tomato fruit-ripening period were determined (Table 4).

Table 2. Tomato fruit quality at different planting periods.

	L*	a*	b*	CI	SSC (g/100 g)	OAC (g/100 g)	SSC/OAC	Vc (mg/100g)	Lycopene (mg/kg)	Fruit Firmness (kg/cm^2)	Comprehensive Quality
T1	42.10 ± 2.00a	14.95 ± 2.61bc	13.69 ± 2.07abc	35.03 ± 2.94bc	3.26 ± 0.23a	0.52 ± 0.1cd	6.32	29.70 ± 1.21bc	21.57 ± 1.94ab	5.98 ± 1.51d	0.865
T2	41.13 ± 3.83ab	11.06 ± 1.33d	12.05 ± 3.61cd	32.87 ± 8.66c	3.19 ± 0.13ab	0.51 ± 0.02d	6.31	29.15 ± 0.90cd	19.97 ± 1.13bc	6.57 ± 1.05d	0.843
T3	41.68 ± 1.42a	16.27 ± 2.15abc	13.37 ± 1.85abc	37.08 ± 3.91abc	3.07 ± 0.12ab	0.60 ± 0.02b	5.12	26.75 ± 0.53d	19.60 ± 0.27c	7.98 ± 0.70bc	0.422
T4	39.24 ± 2.05c	14.21 ± 1.73c	11.28 ± 2.12d	39.93 ± 3.83a	3.10 ± 0.05ab	0.60 ± 0.04b	5.18	23.25 ± 1.91e	15.22 ± 0.29e	9.60 ± 2.30a	0.391
T5	41.67 ± 1.29a	14.91 ± 3.58bc	13.57 ± 2.56abc	35.50 ± 7.86bc	3.00 ± 0.04b	0.71 ± 0.04a	4.22	22.95 ± 1.91e	13.45 ± 2.15f	9.52 ± 1.01a	0.016
T6	41.30 ± 2.08ab	16.74 ± 2.06ab	13.99 ± 2.63abc	37.16 ± 4.85abc	3.17 ± 0.16ab	0.50 ± 0.03d	6.34	32.90 ± 0.76a	22.72 ± 0.65a	6.77 ± 1.04cd	0.895
T7	40.69 ± 1.64abc	17.67 ± 1.93a	14.27 ± 1.39ab	38.24 ± 2.66ab	3.10 ± 0.25ab	0.59 ± 0.02b	5.25	31.20 ± 4.07abc	22.37 ± 0.38a	8.03 ± 1.51bc	0.517
T8	40.72 ± 1.62abc	17.75 ± 2.64a	15.16 ± 2.63a	37.34 ± 2.83abc	3.09 ± 0.04ab	0.52 ± 0.01cd	5.94	32.60 ± 1.13ab	22.33 ± 1.81a	6.61 ± 0.33d	0.830
T9	39.58 ± 1.19bc	15.38 ± 2.46bc	12.91 ± 1.12bcd	38.71 ± 3.01ab	3.00 ± 0.22b	0.53 ± 0.02cd	5.66	32.25 ± 0.57ab	20.08 ± 1.17bc	7.35 ± 0.36bcd	0.749
T10	38.99 ± 1.15c	17.90 ± 1.21a	13.52 ± 1.29abc	40.93 ± 1.63a	3.15 ± 0.06ab	0.54 ± 0.03c	5.82	31.60 ± 1.62abc	17.79 ± 0.90d	8.33 ± 1.23ab	0.666

L* represents black and white, a* represents red and green, and b* represents yellow and blue. CI is color index, SSC is soluble sugar concentration, OAC is organic acid concentration, and Vc is vitamin C. Letters following the values of the indices of each season within rows are the significant differences according to Duncan's multiple range tests at $p < 0.05$; presented values are means ± SD.

Table 3. Weight of tomato fruit quality at red-ripening stage.

Factor	w_{Q1}		w_{Q2}			w_{Q3}	
Sub-Factor	w_{Q11}	w_{Q12}	w_{Q21}	w_{Q22}	w_{Q23}	w_{Q31}	w_{Q32}
	0.158		0.680			0.162	
	0.457	0.543	0.320	0.519	0.161	0.426	0.574
Weight	0.072	0.086	0.218	0.353	0.109	0.069	0.093

Table 4. Optimum range of environmental factors at fruit-ripening stage.

Daily Mean Temperature (°C)	Daytime Mean Temperature (°C)	Night-Time Mean Temperature (°C)	PAR ($\mu mol \cdot m^{-2} \cdot s^{-1}$)	Insolation Duration ($h \cdot d^{-1}$)
18.69 ± 1.35	26.30 ± 2.67	14.30 ± 1.21	592.34 ± 88.74	9.86 ± 0.94

3.2. Development of the Simulation Model for Tomato Quality

3.2.1. Development of Tomato Quality Indices

In the first half of the year, at the red-ripening stage, there was no significant change in color values with the extension of the planting period (from T1 to T5). SSC, Vc, and lycopene slightly declined; OAC and fruit firmness slightly increased (Table 2). In the second half of the year, the tomato fruit quality indicators of five planting times were collected every 5 days from the completion of fruit expansion, including green-ripening stage (GM), veraison stage (V), and red-ripening stage (RR). The fruit quality indices are shown in Table 5. With a delay in planting, the duration tended to be longer in the veraison stage during fruit-ripening periods. a*, SSC, Vc, and lycopene significantly increased from mature green to red-ripening; L*, b*, OAC, and fruit firmness significantly declined. Therefore, the relationship between tomato quality and temperature and light was further investigated based on the data collected in the second half of the year.

Table 5. Quality indices of tomatoes during fruit-ripening period and the corresponding TEP.

Treatments	Stages	T6	T7	T8	T9	T10
TEP	GM	249.70	232.49	248.64	245.40	260.16
	V_1	267.65	247.68	258.81	260.16	266.53
	V_2			269.87	266.53	271.29
	V_3			280.41	271.29	276.58
	RR	283.39	263.97	290.10	276.58	278.11
L*	GM	47.02 ± 1.84bcd	48.83 ± 1.81ab	48.19 ± 1.20bc	46.69 ± 1.06cde	45.58 ± 1.85def
	V_1	45.10 ± 3.58ef	44.38 ± 3.53fg	50.07 ± 2.22a	46.48 ± 2.62cde	47.58 ± 3.26bc
	V_2			47.36 ± 1.98bcd	47.90 ± 1.29bc	43.78 ± 2.23fgh
	V_3			42.19 ± 1.12hij	44.65 ± 1.71f	42.86 ± 1.95ghi
	RR	41.30 ± 2.08ijk	40.69 ± 1.64jkl	40.72 ± 1.62jkl	39.58 ± 1.19kl	38.99 ± 1.15l
a*	GM	−6.54 ± 0.62g	−6.95 ± 0.95g	−6.64 ± 0.43g	−6.49 ± 0.58g	−6.63 ± 0.64g
	V_1	7.30 ± 5.05d	12.14 ± 3.30c	−6.98 ± 0.87g	−6.41 ± 1.12g	−5.17 ± 1.09g
	V_2			−2.29 ± 1.14f	−0.09 ± 2.03e	7.84 ± 1.98d
	V_3			17.73 ± 2.12a	11.77 ± 2.84c	16.58 ± 1.76ab
	RR	16.74 ± 2.06ab	17.67 ± 1.93a	17.75 ± 2.64a	15.38 ± 2.46b	17.90 ± 1.21a
b*	GM	22.51 ± 1.69abc	23.30 ± 1.79ab	23.19 ± 2.15ab	20.65 ± 2.39cde	20.35 ± 2.24de
	V_1	19.23 ± 2.37e	16.34 ± 2.94f	23.73 ± 1.96a	20.73 ± 2.81cde	21.64 ± 2.44bcd
	V_2			24.53 ± 3.41a	23.26 ± 3.22ab	15.38 ± 1.64fg
	V_3			14.74 ± 1.61fgh	14.51 ± 1.80fgh	14.27 ± 1.95gh
	RR	13.99 ± 2.63gh	14.27 + 1.39gh	15.16 ± 2.63fg	12.91 ± 1.12h	13.52 ± 1.29gh
CI	GM	−11.94 ± 1.75hi	−11.77 ± 0.92hi	−11.51 ± 8.48hi	−12.91 ± 3.38hi	−13.69 ± 4.85i
	V_1	15.23 ± 1.80e	27.27 ± 1.62c	−11.29 ± 1.11hi	−12.93 ± 4.04hi	−9.84 ± 2.66h
	V_2			−3.89 ± 2.80g	−0.40 ± 3.54f	20.49 ± 2.83d
	V_3			36.34 ± 2.05b	27.87 ± 4.60c	35.42 ± 3.01b
	RR	37.16 ± 1.39b	38.24 ± 9.84ab	37.34 ± 1.78b	38.71 ± 3.35ab	40.93 ± 1.63a

Table 5. *Cont.*

Treatments	Stages	T6	T7	T8	T9	T10
SSC (g/100 g)	GM	2.60 ± 0.12f	2.57 ± 0.24f	2.59 ± 0.12f	2.61 ± 0.10ef	2.85 ± 0.26bcdef
	V_1	2.86 ± 0.18abcde	2.80 ± 0.03bcdef	2.69 ± 0.12def	2.81 ± 0.20bcdef	2.93 ± 0.14abcd
	V_2			2.79 ± 0.16cdef	2.84 ± 0.22bcdef	2.95 ± 0.18abcd
	V_3			3.05 ± 0.09abc	2.85 ± 0.11bcdef	3.06 ± 0.34abc
	RR	3.17 ± 0.16a	3.10 ± 0.25ab	3.09 ± 0.04abc	3.00 ± 0.22abc	3.15 ± 0.06a
OAC (g/100 g)	GM	0.69 ± 0.05cde	0.80 ± 0.01a	0.78 ± 0.03ab	0.73 ± 0.13bc	0.71 ± 0.02cd
	V_1	0.64 ± 0.03efg	0.67 ± 0.04de	0.70 ± 0.02cd	0.59 ± 0.03fghi	0.65 ± 0.01def
	V_2			0.60 ± 0.03fgh	0.55 ± 0.01hijk	0.60 ± 0.03fgh
	V_3			0.56 ± 0.02hij	0.54 ± 0.01ijk	0.57 ± 0.03hij
	RR	0.50 ± 0.03k	0.59 ± 0.02ghi	0.52 ± 0.01jk	0.53 ± 0.02jk	0.54 ± 0.03hijk
SSC/OAC	GM	3.76	3.21	3.32	3.56	4.01
	V_1	4.49	4.20	3.85	4.72	4.51
	V_2			4.65	5.16	4.93
	V_3			5.44	5.30	5.39
	RR	6.34	5.25	5.94	5.66	5.82
Vc (mg/100 g)	GM	16.15 ± 0.53e	16.95 ± 0.50e	16.85 ± 0.53e	15.30 ± 1.64e	17.35 ± 0.90e
	V_1	26.25 ± 0.30bc	24.00 ± 1.23c	20.85 ± 3.92d	20.05 ± 0.62d	26.10 ± 0.82bc
	V_2			25.90 ± 0.77bc	28.30 ± 2.61b	28.15 ± 1.62b
	V_3			31.65 ± 0.25a	31.65 ± 0.82a	31.60 ± 1.40a
	RR	32.90 ± 0.76a	31.20 ± 4.07a	32.60 ± 1.13a	32.25 ± 0.57a	31.60 ± 1.62a
Lycopene (mg/kg)	GM	4.40 ± 0.22i	4.18 ± 0.22i	6.75 ± 0.46h	7.32 ± 0.13gh	6.84 ± 0.53h
	V_1	9.90 ± 2.91ef	11.44 ± 0.90e	6.93 ± 0.58h	7.23 ± 0.11gh	7.93 ± 0.52gh
	V_2			10.90 ± 0.56e	8.96 ± 0.81fg	10.26 ± 0.32ef
	V_3			21.89 ± 3.80a	16.37 ± 0.26c	13.78 ± 0.58d
	RR	22.72 ± 0.65a	22.37 ± 0.38a	22.32 ± 1.81a	20.08 ± 1.17b	17.79 ± 0.90c
Fruit firmness (kg/cm^2)	GM	13.53 ± 1.50bc	15.39 ± 0.92a	15.03 ± 0.27a	14.99 ± 0.53a	14.64 ± 0.77ab
	V_1	10.18 ± 1.12e	11.57 ± 0.90d	13.68 ± 0.49bc	13.53 ± 0.59bc	13.57 ± 1.05bc
	V_2			8.76 ± 1.14fg	13.10 ± 0.59c	12.46 ± 1.47cd
	V_3			7.45 ± 0.98hi	8.78 ± 1.08fg	9.82 ± 2.83ef
	RR	6.77 ± 1.04j	8.03 ± 1.51gh	6.60 ± 0.33ij	7.35 ± 0.36hi	8.33 ± 1.23gh

Notes: GM, V, and RR represent the green mature stage, the veraison stage, and the red-ripening stage, respectively. Different letters represent significant differences at $p < 0.05$ level.

We conducted a correlation analysis between the tomato quality and TEP, which was calculated using Equations (1)–(4). As shown in Table 6, there was a positive correlation between TEP and SSC/OAC (r = 0.917), SSC (r = 0.869), lycopene (r = 0.774), and Vc (r = 0.869). TEP was negative correlated with OAC (r = −0.897), and there were high and moderate correlations between TEP and fruit firmness (r = −0.823) and CI (r = 0.709). The effects of TEP on SSC, OAC, the sugar–acid ratio, and Vc were stronger than those of lycopene and fruit firmness. The SSC, OAC, and SSC/OAC of tomato fruit are the main components of taste quality and the main source of the tomato commodity value. Therefore a tomato quality (SSC, OAC, and SSC/OAC) simulation model was constructed based on TEP to predict the effects of temperature and light on the tomato quality.

Table 6. Correlation coefficients (r) for tomato fruit quality and TEP.

	CI	SSC	OAC	SSC/OAC	Lycopene	Vc	Fruit Firmness
TEP	0.709 **	0.869 **	−0.897 **	0.917 **	0.774 **	0.869 **	−0.823 **

Notes: ** indicates significant levels at $p < 0.01$.

3.2.2. Development of the Simulation Model for Tomato Fruit Single-Quality Indices

We established a simulation model based on TEP for the quality of tomatoes during fruit-ripening periods that included SSC, OAC, and SSC/OAC (Table 7). TEP was obtained using Formulas (1)–(4) to process the environmental data. The analysis showed that the relationship between SSC and TEP was consistent with the change in the logarithmic function curve and R^2 was 0.750. The relationship between OAC and TEP conformed to the first-order function and R^2 was 0.808. The sugar–acid ratio changed with TEP in the logarithmic function curve and R^2 was 0.833.

Table 7. Simulation model of tomato fruit quality index based on TEP.

Fruit Quality	Model Equations	a	b	R^2	RMSE
SSC	$SSC = a\ln X + b$	0.031	0.139	0.750	0.09%
OAC	$OAC = aX + b$	-5×10^{-5}	0.021	0.808	0.14%
SSC/OAC	$y = a\ln X + b$	14.989	−78.874	0.833	0.358

Note: X is TEP (MJ/m^2); a and b are parameters of the equation. RMSE is the root mean square error.

In order to determine the accuracy of the model, the simulated values were calculated using the equation and compared with the measured values by calculating the RMSE. The results showed that the RSME value of the simulated models for SSC was 0.09%; this indicated that the model had high simulation accuracy and the variation trend of the simulated and measured values was the same (Figure 3A). The OAC simulation value had a large deviation from the measured value and the RMSE value was higher (0.14%), indicating that the model had moderate simulation accuracy (Figure 3B). The RSME value of the simulated models for the sugar–acid ratio was 0.358; this indicated that the model had high simulation accuracy (Figure 3C).

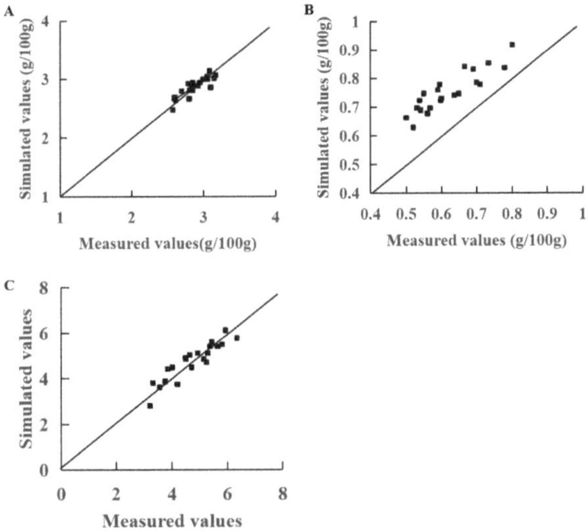

Figure 3. Comparison of simulated and measured values of tomato fruit quality. (**A**) SSC model validation. (**B**) OAC model validation. (**C**) SSC/OAC model validation.

3.2.3. Development of the Simulation Model for Tomato Comprehensive-Quality Indices

Tomato fruit quality is a comprehensive concept that is the result of the joint action of various quality indicators [33]. In order to evaluate tomato fruit quality using TOPSIS, the weight of factor (w_i) in the first layer and the weight of sub-factor (w_{ij}) in the second layer were calculated using AHP and the entropy method, respectively. These two sets of weights were then merged together (Table 8).

Table 8. Weight of tomato fruit quality during the fruit-ripening periods.

Factor	w_{Q1}		w_{Q2}			w_{Q3}	
Sub-Factor	w_{Q11}	w_{Q12}	w_{Q21}	w_{Q22}	w_{Q23}	w_{Q31}	w_{Q32}
	0.158		0.680			0.162	
	0.628	0.372	0.335	0.374	0.290	0.567	0.432
Weight	0.099	0.059	0.228	0.254	0.197	0.092	0.070

After obtaining the comprehensive weight of each quality index of tomato frui TOPSIS was used to calculate the comprehensive value of tomato fruit quality durin the fruit-ripening periods (Table 9). The higher the comprehensive evaluation value, th higher the quality, indicating that the environmental parameters of the treatment wer more conducive to obtaining high-quality tomatoes.

Table 9. The tomato comprehensive-quality indices based on TOPSIS.

	The Green Mature Stage		The Veraison Stage		The Red-Ripening Stage
T6	0.120	0.458			0.953
T7	0.028	0.531			0.799
T8	0.065	0.145	0.346	0.831	0.920
T9	0.098	0.275	0.409	0.720	0.870
T10	0.151	0.254	0.546	0.733	0.852

The simulation model was obtained by analyzing the relationship between the com prehensive evaluation value of fruit and TEP as follows ($R^2 = 0.757$):

$$y = 1.189/(1 + \exp(-0.036(TEP - 271.359))) \quad (17$$

where y is the comprehensive fruit quality evaluation value and 1.189, 0.036, and 271.35 are the parameters of the equation.

The RMSE value was 0.154, which showed that the simulation model provide medium precision (Figure 4).

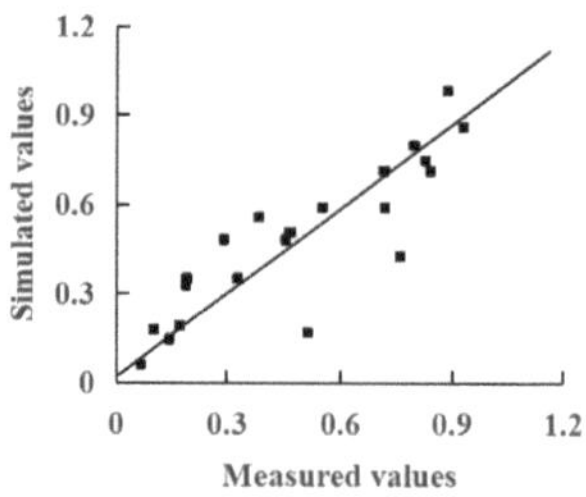

Figure 4. Comparison of simulated and measured comprehensive quality values of tomato fruit.

4. Discussion

With the progress of tomato-planting technology, the yield of tomatoes has greatl improved. In recent years, high-quality tomatoes have gradually become the main target o market demand. In greenhouse cultivation, the environment is a key factor influencing th formation of tomato fruit quality and clarifying the relationship between environmenta factors and the formation of quality is fundamental to achieve high-quality, high-yiel tomatoes. Mathematical modeling has been used to describe the characteristics of cro growth. Establishing a growth model can help to better understand the responses of crop to their environment and improve the efficiency of agricultural production [46]. In thi study, the relationship between fruit quality and environmental factors was studied.

A number of studies have shown that temperature and light stress can reduce qualit formation in tomatoes, including high and low temperatures as well as strong and wea light [13–15,22]. However, a greenhouse environment is a multi-variable, highly couple and complex system. The regulation of environmental factors differs from the superpositio of single factors [32]. There were significant differences in the quality of tomato frui between the different planting periods (Table 2). This indicates that there is an associatio between environmental factors and tomato quality formation.

Previous studies have focused on analyzing the relationship between environmenta factors and crop growth, and then have developed models to effectively predict cro growth, yield, and the duration of growth stage under different environmental conditions such as temperature and light, providing decision-making support for the environmental management of crop growth [28,30,47–49]. For example, a leaf area model was established based on the TEP method that uncovered the difference in plant growth caused by different light/dark cycle patterns from a physiological perspective [47]. However, there are fewer studies on the relationship between tomato fruit quality and environmental factors. In addition, previous crop simulation models mostly adopted single environmental factors such as GDD. However, GDD can only describe the effect of the lower growth-limit temperature and the upper growth-limit temperature on crop growth and development; it does not involve the effects of the photoperiod or high temperature on the retardation of development. Therefore, GDD is more suitable for field crops because the changes in field temperature and solar radiation are basically synchronous. In a cultivation facility, changes in environmental factors are not always synchronized due to heating, cooling, and the uncovering of insulation, so a simulation is less accurate when using a single environmental factor [50,51]. It is necessary to use a comprehensive index based on light and temperature to establish a simulation model [30,52]. The TEP method takes into account both air temperature and light, and can be used to build effective, simple models. In this study, based on the results of a correlation analysis, we established a simulation model for the fruit quality response based on TEP. The results showed that the SSC, OAC, and SSC/OAC of fruit quality in the fruit-ripening period were highly significantly correlated with TEP and the simulation accuracy was high (Table 7; Figure 3).

According to the model results of the different indicators we established (Table 5), it was found that the change trend of different quality indices was not consistent, indicating that it was difficult to achieve the optimal conditions of each index by adjusting the environmental factors. He et al. (2022) adopted a design of composite quadratic orthogonal regressive rotation with three factors and five levels, including temperature, relative humidity (RH), and photosynthetically active radiation (PAR), and established the response models of growth indicators (seedling index, dry-matter accumulation, net photosynthetic rate, transpiration rate, and chlorophyll content) for multiple environmental factors. The results showed that the most suitable environmental parameters of the five indicators were not consistent. It is clear that no single indicator can effectively reflect the growth of plants or the quality of fruit, so a comprehensive evaluation analysis is necessary. Here, the subjective and objective weights of all the quality indices were obtained based on AHP and the entropy method. Combining human subjective judgement and the objective information from index data can reflect the comprehensive quality of tomatoes during the fruit-ripening period in a more reasonable way [32,33,40]. This study comprehensively evaluated the fruit quality of tomatoes at different planting periods during the fruit-ripening period. The results showed that the optimal daily mean temperature during fruit ripening was 18.69 ± 1.35 °C, the daytime mean temperature was 26.30 ± 2.67 °C, the night-time mean temperature was 14.30 ± 1.21 °C, and PAR was 592.34 ± 88.74 $\mu mol \cdot m^{-2} \cdot s^{-1}$, with a sunshine duration of 9.86 ± 0.94 $h \cdot d^{-1}$ (Table 4). Then, we established a simulation model of comprehensive tomato fruit quality based on TEP using a regression analysis. The results showed that the comprehensive-quality formation model of tomato fruit at a mature stage based on TEP was moderately accurate (R^2 = 0.757).

5. Conclusions

This study was carried out in the natural environmental conditions of a greenhouse. In total, 10 batches were grown within the range of environments in which tomatoes can be effectively grown, covering the environmental conditions that may be faced during the cultivation of tomatoes. Based on the correlation analysis between tomato fruit quality and environmental factors and the results of the integrated weighting of each quality index, the main quality indicators, including SSC, OAC, and SSC/OAC, were selected. A model was established to investigate the relationships between SSC, OAC, and SSC/OAC in tomato fruit and TEP during the fruit-ripening period in a solar greenhouse. The model was validated and the results indicated that the model had a high level of simulation accuracy. At the same time, based on a comprehensive evaluation analysis, a TEP-based comprehensive fruit quality formation model was established. The model was validated and the results indicated that the model had moderate simulation accuracy. This study provides decision-making support for the management of temperature and light in a heliostat during tomato ripening.

Supplementary Materials: The following supporting information can be downloaded at: https://www.mdpi.com/article/10.3390/plants13121662/s1.

Author Contributions: F.Y. and J.L.: Conceptualization, Methodology, Supervision, Writing—Review and Editing, and Funding Acquisition. Y.Q.: Data Curation, Investigation, and Writing—Original Draft. A.G.: Investigation, Formal Analysis, and Software. X.L. and N.L.: Investigation and Methodology. T.J.: Methodology and Writing—Review and Editing. All authors have read and agreed to the published version of the manuscript.

Funding: The study was supported by the National Key Research and Development Program (2023YFD2300702), the Major Scientific Innovation Project of Shandong Province (2019JZZY010715), the Shandong Provincial Key Research and Development Program (2022LZGC009), and the Key Research and Development Program of Ningxia Hui Autonomous Region (2023BCF01042).

Data Availability Statement: Data are contained within the article and Supplementary Materials.

Conflicts of Interest: The authors declare no conflicts of interest.

References

1. Mesa, T.; Polo, J.; Arabia, A.; Caselles, V.; Munne-Bosch, S. Differential physiological response to heat and cold stress of tomato plants and its implication on fruit quality. *J. Plant Physiol.* **2022**, *268*, 153581. [CrossRef] [PubMed]
2. Li, J.M.; Xiang, C.Y.; Wang, X.X.; Guo, Y.M.; Huang, Z.J.; Liu, L.; Li, X.; Du, Y.C. Current situation of tomato industry in china during 'The Thirteenth Five-year Plan' period and future prospect. *China Veg.* **2021**, *2*, 13–20. [CrossRef]
3. Erba, D.; Casiraghi, M.C.; Ribas-Agustí, A.; Cáceres, R.; Marfà, O.; Castellari, M. Nutritional value of tomatoes (*Solanum lycopersicum* L.) grown in greenhouse by different agronomic techniques. *J. Food Compos. Anal.* **2013**, *31*, 245–251. [CrossRef]
4. Massot, C.; Genard, M.; Stevens, R.; Gautier, H. Fluctuations in sugar content are not determinant in explaining variations in vitamin C in tomato fruit. *Plant Physiol. Biochem.* **2010**, *48*, 751–757. [CrossRef] [PubMed]
5. Ashraf, M.; Foolad, M.R. Roles of glycine betaine and proline in improving plant abiotic stress resistance. *Environ. Exp. Bot.* **2007**, *59*, 206–216. [CrossRef]
6. Ummenhofer, C.C.; Meehl, G.A. Extreme weather and climate events with ecological relevance: A review. *Philos. Trans. R. Soc. Lond. B Biol. Sci.* **2017**, *372*, 20160135. [CrossRef] [PubMed]
7. Adams, S.R.; Cockshull, K.E.; Cave, C.R.J. Effect of temperature on the growth and development of tomato fruits. *Ann. Bot.* **2001**, *88*, 869–877. [CrossRef]
8. Alsamir, M.; Mahmood, T.; Trethowan, R.; Ahmad, N. An overview of heat stress in tomato (*Solanum lycopersicum* L.). *Saudi J. Biol. Sci.* **2021**, *28*, 1654–1663. [CrossRef] [PubMed]
9. Lozano, R.; Angosto, T.; Gómez, P.; Payán, C.; Capel, J.; Huijser, P.; Salinas, J.; Martınez-Zapater, J.M. Tomato flower abnormalities induced by low temperatures are associated with changes of expression of MADS-Box genes1. *Plant Physiol.* **1998**, *117*, 91–100. [CrossRef]
10. Meena, Y.K.; Khurana, D.S.; Kaur, N.; Singh, K. Towards enhanced low temperature stress tolerance in tomato: An approach. *J. Environ. Biol.* **2018**, *39*, 529–535. [CrossRef]
11. Truffault, V.; Fifel, F.; Longuenesse, J.-J.; Vercambre, G.; Le Quillec, S.; Gautier, H. Impact of temperature integration under greenhouse on energy use efficiency, plant growth and development and tomato fruit quality depending on cultivar rootstock combination. *II Int. Symp. Hortic. Eur.* **2012**, *1099*, 95–100. [CrossRef]
12. Hatfield, J.L.; Prueger, J.H. Temperature extremes: Effect on plant growth and development. *Weather Clim. Extrem.* **2015**, *10*, 4–10. [CrossRef]
13. Rosales, M.A.; Cervilla, L.M.; Sanchez-Rodriguez, E.; Rubio-Wilhelmi, M.D.; Blasco, B.; Rios, J.J.; Soriano, T.; Castilla, N.; Romero, L.; Ruiz, J.M. The effect of environmental conditions on nutritional quality of cherry tomato fruits: Evaluation of two experimental Mediterranean greenhouses. *J. Sci. Food Agric.* **2011**, *91*, 152–162. [CrossRef] [PubMed]
14. Dumas, Y.; Dadomo, M.; Di Lucca, G.; Grolier, P. Effects of environmental factors and agricultural techniques on antioxidantcontent of tomatoes. *J. Sci. Food Agric.* **2003**, *83*, 369–382. [CrossRef]
15. Vijayakumar, A.; Shaji, S.; Beena, R.; Sarada, S.; Rani, T.S.; Stephen, R.; Manju, R.V.; Viji, M.M. High temperature induced changes in quality and yield parameters of tomato (*Solanum lycopersicum* L.) and similarity coefficients among genotypes using SSR markers. *Heliyon* **2021**, *7*, e05988. [CrossRef] [PubMed]

16. Dannehl, D.; Huber, C.; Rocksch, T.; Huyskens-Keil, S.; Schmidt, U. Interactions between changing climate conditions in a semi-closed greenhouse and plant development, fruit yield, and health-promoting plant compounds of tomatoes. *Sci. Hortic.* **2012**, *138*, 235–243. [CrossRef]
17. Hernández, V.; Hellín, P.; Fenoll, J.; Molina, M.V.; Garrido, I.; Flores, P. Impact of high temperature stress on ascorbic acid concentration in tomato. *VIII Int. Postharvest Symp. Enhancing Supply Chain. Consum. Benefits-Ethical Technol. Issues* **2016**, *1194*, 985–990. [CrossRef]
18. Petridis, A.; van der Kaay, J.; Chrysanthou, E.; McCallum, S.; Graham, J.; Hancock, R.D. Photosynthetic limitation as a factor influencing yield in highbush blueberries (Vaccinium corymbosum) grown in a northern European environment. *J. Exp. Bot.* **2018**, *69*, 3069–3080. [CrossRef] [PubMed]
19. Aliniaeifard, S.; van Meeteren, U. Stomatal characteristics and desiccation response of leaves of cut chrysanthemum (*Chrysanthemum morifolium*) flowers grown at high air humidity. *Sci. Hortic.* **2016**, *205*, 84–89. [CrossRef]
20. Arve, L.E.; Kruse, O.M.; Tanino, K.K.; Olsen, J.E.; Futsaether, C.; Torre, S. Daily changes in VPD during leaf development in high air humidity increase the stomatal responsiveness to darkness and dry air. *J. Plant Physiol.* **2017**, *211*, 63–69. [CrossRef]
21. Mei, Z.; Li, Z.; Lu, X.; Zhang, S.; Liu, W.; Zou, Q.; Yu, L.; Fang, H.; Zhang, Z.; Mao, Z.J.E.; et al. Supplementation of natural light duration promotes accumulation of sugar and anthocyanins in apple (*Malus domestica* Borkh.) fruit. *Environ. Exp. Bot.* **2023**, *205*, 105133. [CrossRef]
22. Demers, D.-A.; Dorais, M.; Wien, C.H.; Gosselin, A.J.S.H. Effects of supplemental light duration on greenhouse tomato (*Lycopersicon esculentum* Mill.) plants and fruit yields. *Sci. Hortic.* **1998**, *74*, 295–306. [CrossRef]
23. Zou, L.M.; Zhong, G.Y.; Wu, B.H.; Yang, Y.Z.; Li, S.H.; Liang, Z.C. Effects of sunlight on anthocyanin accumulation and associated co-expression gene networks in developing grape berries. *Environ. Exp. Bot.* **2019**, *166*, 103811. [CrossRef]
24. Li, Y.; Hu, J.; Wei, H.; Jeong, B.R. A Long-Day Photoperiod and 6-Benzyladenine Promote Runner Formation through Upregulation of Soluble Sugar Content in Strawberry. *Int. J. Mol. Sci.* **2020**, *21*, 4917. [CrossRef] [PubMed]
25. Franklin, K.A.; Toledo-Ortiz, G.; Pyott, D.E.; Halliday, K.J. Interaction of light and temperature signaling. *J. Exp. Bot.* **2014**, *65*, 2859–2871. [CrossRef] [PubMed]
26. Gerganova, M.; Popova, A.V.; Stanoeva, D.; Velitchkova, M. Tomato plants acclimate better to elevated temperature and high light than to treatment with each factor separately. *Plant Physiol. Bioch.* **2016**, *104*, 234–241. [CrossRef] [PubMed]
27. Wei, Z.H.; Du, T.S.; Li, X.N.; Fang, L.; Liu, F.L. Interactive effects of elevated CO_2 and N fertilization on yield and quality of tomato grown under reduced irrigation regimes. *Front. Plant Sci.* **2018**, *9*, 326830. [CrossRef] [PubMed]
28. Doan, C.C.; Tanaka, M. Relationships between Tomato Cluster Growth Indices and Cumulative Environmental Factors during Greenhouse Cultivation. *Sci. Hortic.* **2018**, *295*, 110803. [CrossRef]
29. Wu, Y.; Yan, S.C.; Fan, J.L.; Zhang, F.C.; Xiang, Y.Z.; Zheng, J.; Guo, J.J. Responses of growth, fruit yield, quality and water productivity of greenhouse tomato to deficit drip irrigation. *Sci. Hortic.* **2021**, *275*, 109710. [CrossRef]
30. Zhou, T.M.; Wu, Z.; Wang, Y.C.; Su, X.J.; Qin, C.X.; Huo, H.Q.; Jiang, F.L. Modelling seedling development using thermal effectiveness and photosynthetically active radiation. *J. Integr. Agric.* **2019**, *18*, 2521–2533. [CrossRef]
31. Chen, D.Y.; Zhang, J.H.; Zhang, B.; Wang, Z.S.; Xing, L.B.; Zhang, H.H.; Hu, J. Obtaining a light intensity regulation target value based on the tomato dry weight model. *Sci. Hortic.* **2022**, *295*, 110879. [CrossRef]
32. He, Z.H.; Su, C.J.; Cai, Z.L.; Wang, Z.; Li, R.; Liu, J.C.; He, J.Q.; Zhang, Z. Multi-factor coupling regulation of greenhouse environment based on comprehensive growth of cherry tomato seedlings. *Sci. Hortic.* **2022**, *297*, 110960. [CrossRef]
33. Wang, F.; Kang, S.Z.; Du, T.S.; Li, F.S.; Qiu, R.J. Determination of comprehensive quality index for tomato and its response to different irrigation treatments. *Agric. Water Manag.* **2011**, *98*, 1228–1238. [CrossRef]
34. Wang, C.X.; Gu, F.; Chen, J.L.; Yang, H.; Jiang, J.J.; Du, T.S.; Zhang, J.H. Assessing the response of yield and comprehensive fruit quality of tomato grown in greenhouse to deficit irrigation and nitrogen application strategies. *Agric. Water Manag.* **2015**, *161*, 9–19. [CrossRef]
35. Ripoll, J.; Urban, L.; Staudt, M.; Lopez-Lauri, F.; Bidel, L.P.R.; Bertin, N. Water shortage and quality of fleshy fruits-making the most of the unavoidable. *J. Exp. Bot.* **2014**, *65*, 4097–4117. [CrossRef]
36. Marti, R.; Rosello, S.; Cebolla-Cornejo, J. Tomato as a source of carotenoids and polyphenols targeted to cancer prevention. *Cancers* **2016**, *8*, 58. [CrossRef]
37. Araya, N.A.; Chiloane, T.S.; Rakuambo, J.Z.; Maboko, M.M.; du Plooy, C.P.; Amoo, S.O. Effect of environmental variability on fruit quality and phytochemical content of soilless grown tomato cultivars in a non-temperature-controlled high tunnel. *Sci. Hortic.* **2021**, *288*, 110378. [CrossRef]
38. Zhou, D.X.; Su, Y.; Ning, Y.C.; Rong, G.H.; Wang, G.D.; Liu, D.; Liu, L.Y. Estimation of the Effects of Maize Straw Return on Soil Carbon and Nutrients Using Response Surface Methodology. *Pedosphere* **2018**, *28*, 411–421. [CrossRef]
39. Jiang, X.L.; Zhao, Y.L.; Tong, L.; Wang, R.; Zhao, S. Quantitative analysis of tomato yield and comprehensive fruit quality in response to deficit irrigation at different growth stages. *Hortscience* **2019**, *54*, 1409–1417. [CrossRef]
40. He, Z.H.; Li, M.N.; Cai, Z.L.; Zhao, R.S.; Hong, T.T.; Yang, Z.; Zhang, Z. Optimal irrigation and fertilizer amounts based on multi-level fuzzy comprehensive evaluation of yield, growth and fruit quality on cherry tomato. *Agric. Water Manag.* **2020**, *243*, 106360. [CrossRef]
41. Tian, Y.N.; Cao, L.L.; Zhao, L.Q.; Cao, C.H. Effect of different nutrient solution formulas on quality of tomato seedling in tidal irrigation. *China Fruit Veg.* **2021**, *41*, 83–87. [CrossRef]

42. Teng, L.; Cheng, Z.; Chen, X.; Lai, L. Study on simulation models of tomato fruit quality related to cultivation environmenta factors. *Acta Ecol. Sin.* **2021**, *32*, 111–116. [CrossRef]
43. Shao, Y.Y.; Shi, Y.K.; Qin, Y.D.; Xuan, G.T.; Li, J.; Li, Q.K.; Yang, F.J.; Hu, Z.C. A new quantitative index for the assessment c tomato quality using Vis-NIR hyperspectral imaging. *Food Chem.* **2022**, *386*, 132864. [CrossRef]
44. Kundu, S.; Khare, D.; Mondal, A. Landuse change impact on sub-watersheds prioritization by analytical hierarchy process (AHP *Ecol. Inform.* **2017**, *42*, 100–113. [CrossRef]
45. Fleming, A.; Schenkel, F.S.; Chen, J.; Malchiodi, F.; Bonfatti, V.; Ali, R.A.; Mallard, B.; Corredig, M.; Miglior, F. Prediction of mil fatty acid content with mid-infrared spectroscopy in Canadian dairy cattle using differently distributed model development set *J. Dairy Sci.* **2017**, *100*, 5073–5081. [CrossRef]
46. Fang, S.-L.; Kuo, Y.-H.; Kang, L.; Chen, C.-C.; Hsieh, C.-Y.; Yao, M.-H.; Kuo, B.-J. Using Sigmoid Growth Models to Simulat Greenhouse Tomato Growth and Development. *Horticulturae* **2022**, *8*, 1021. [CrossRef]
47. Hang, T.; Lu, N.; Takagaki, M.; Mao, H.P. Leaf area model based on thermal effectiveness and photosynthetically active radiatio in lettuce grown in mini-plant factories under different light cycles. *Sci. Hortic.* **2019**, *252*, 113–120. [CrossRef]
48. Ding, D.Y.; Feng, H.; Zhao, Y.; Hill, R.L.; Yan, H.M.; Chen, H.X.; Hou, H.J.; Chu, X.S.; Liu, J.C.; Wang, N.J.; et al. Effects c continuous plastic mulching on crop growth in a winter wheat-summer maize rotation system on the Loess Plateau of Chin *Agric. Forest Meteorol.* **2019**, *271*, 385–397. [CrossRef]
49. Jia, B.; He, H.B.; Ma, F.Y.; Diao, M.; Jiang, G.Y.; Zheng, Z.; Cui, J.; Fan, H. Modeling aboveground biomass accumulation of cotto *J. Anim. Plant Sci.* **2014**, *24*, 280–289. [CrossRef]
50. Wang, D.; Li, G.-r.; Zhou, B.-y.; Zhan, M.; Cao, C.-g.; Meng, Q.-f.; Xia, F.; Ma, W.; Zhao, M. Innovation of the double-maiz cropping system based on cultivar growing degree days for adapting to changing weather conditions in the North China Plai *J. Integr. Agric.* **2020**, *19*, 2997–3012. [CrossRef]
51. Liu, Y.; Su, L.; Wang, Q.; Zhang, J.; Shan, Y.; Deng, M. Comprehensive and quantitative analysis of growth characteristics o winter wheat in China based on growing degree days. *Adv. Agron.* **2020**, *159*, 237–273. [CrossRef]
52. Xu, R.; Dai, J.; Luo, W.; Yin, X.; Li, Y.; Tai, X.; Han, L.; Chen, Y.; Lin, L.; Li, G. A photothermal model of leaf area index fo greenhouse crops. *Agric. Forest. Meteorol.* **2010**, *150*, 541–552. [CrossRef]

plants

Article

An Integrated Analysis of Anatomical and Sugar Contents Identifies How Night Temperatures Regulate the Healing Process of Oriental Melon Grafted onto Pumpkin

Huan Liang, Jiangfeng Liu, Xianfeng Shi, Mihong Ge, Juhong Zhu, Dehuan Wang and Mobing Zhou *

Wuhan Academy of Agricultural Sciences, Wuhan 430070, China; lianghuanconf@126.com (H.L.); ljf980223@163.com (J.L.); shixf124@163.com (X.S.); gmh917@126.com (M.G.); hongye408@163.com (J.Z.); wdhuan1987@163.com (D.W.)
* Correspondence: zmb19682022@163.com; Tel.: +86-027-6230-3086

Abstract: Graft healing is a complex process affected by environmental factors, with temperature being one of the most important influencing factors. Here, oriental melon grafted onto pumpkin was used to study changes in graft union formation and sugar contents at the graft interface under night temperatures of 18 °C and 28 °C. Histological analysis suggested that callus formation occurred 3 days after grafting with a night temperature of 28 °C, which was one day earlier than with a night temperature of 18 °C. Vascular reconnection with a night temperature of 28 °C was established 2 days earlier than with a night temperature of 18 °C. Additionally, nine sugars were significantly enriched in the graft union, with the contents of sucrose, trehalose, raffinose, D–glucose, D–fructose, D–galactose, and inositol initially increasing but then decreasing. Furthermore, we also found that exogenous glucose and fructose application promotes vascular reconnection. However, exogenous sucrose application did not promote vascular reconnection. Taken together, our results reveal that elevated temperatures improve the process of graft union formation through increasing the contents of sugars. This study provides information to develop strategies for improving grafting efficiency under low temperatures.

Keywords: temperature; melon; healing process; anatomical; sugar

check for updates

Citation: Liang, H.; Liu, J.; Shi, X.; Ge, M.; Zhu, J.; Wang, D.; Zhou, M. An Integrated Analysis of Anatomical and Sugar Contents Identifies How Night Temperatures Regulate the Healing Process of Oriental Melon Grafted onto Pumpkin. *Plants* **2024**, 13, 1506. https://doi.org/10.3390/plants13111506

Academic Editor: Christopher Cazzonelli

Received: 21 April 2024
Revised: 22 May 2024
Accepted: 29 May 2024
Published: 30 May 2024

1. Introduction

In vegetable crops, grafting has been commonly used for counteracting adverse cultivation situations, including temperature changes, organic pollutants, heavy metals, high salinity, drought, insect pests, and soil-borne pathogens [1,2]. Grafting is considered to be a cost-effective and environmentally friendly operation. In addition, this technique improves fruit yield and productivity by enhancing water use efficiency and nutrient uptake [3]. Worldwide, a high percentage of grafted *Cucurbitaceae* and *Solanaceae* vegetables has been widely applied in agricultural practice [4]. However, environmental stress at the graft union formation stage leads to a low survival rate.

It is widely recognized that the successful healing of rootstocks and scions is a crucial prerequisite for the cultivation of grafted seedlings. Successful grafting, involving the initial adhesion of scion and rootstock, callus tissue formation, and vascular reconnection, depends on internal factors, including hormones, sugars, and the developmental stage, but also on external factors, such as temperature, humidity, light, and culture methods [5–7]. Optimizing environmental factors during the healing and acclimatization stages is essential to produce high-quality grafted seedlings. One notable factor is temperature, which influences the growth rate of the plant, the rate of wound healing, and regeneration. In Walnut, the graft success rate increased from 6% to 73% with localized heating [8]. In Arabidopsis, raising the healing temperature from 22 °C to 27 °C speed up grafting and vascular reconnection by approximately 25% [9]. In tomato, the tensile strength of the graft union was greatly improved when the stored temperature of the graft union ranged from 23 °C to 27 °C [10].

Melon (*Cucumis melo* L.) is an important horticultural crop that belongs to the *Cucurbitaceae* family. In China, melon plants are mainly grafted during in early spring, when plants often suffer from low temperatures. The night healing temperature decreases to 12 °C, which results in a wider gully between the rootstock and scion cut surface and delays the differentiation of vascular tissue, leading to unsuccessful graft union formation [11]. The optimal healing temperature ranges from 22 °C to 28 °C [12,13]. In agricultural practice, heating is necessary to ensure the high survival rate of the grafted seedlings. However, heating is interlinked with energy consumption and has economic impacts. So, understanding how temperature regulates the biological processes during graft union formation under low-temperature conditions is crucial to achieve a sustainable agricultural system.

Sugars are active and essential during graft union formation [14]. Sugars, in addition to their fundamental roles as carbon and energy sources, also act as signaling molecules to regulate gene expression. Almansa [15] found that the graft healing ability of macadamia nuts was influenced by the carbohydrate content of the scion. Marsch-Martínez [16] found that adding 0.5% sugar to the grafting medium resulted in faster recovery after grafting and a higher graft success rate compared to plants with 0% sugar. However, the function of carbohydrates in graft union formation under low temperatures has not yet been studied.

To reduce the impact of external environmental factors and improve the healing efficiency of melon grafted plug seedlings, we performed an anatomical analysis and integrated metabolic analysis to observe the cellular and metabolite changes that occur during graft union formation. Furthermore, the application of exogenous glucose, fructose or sucrose demonstrated that glucose and fructose play important roles in melon graft union formation under low night temperatures. These results provide information to develop strategies for improving grafting techniques, as well as scientific instruction for the sensible and efficient management of grafted melon seedlings.

2. Results

2.1. Anatomical Observation during Graft Union Formation

Graft healing formation was divided into the following three recognizable developmental stages: the isolated layer (IL) stage, the callus (CA) stage, and the vascular bundles (VBs) stage (Figure 1A,D). To investigate the effect of night temperature treatment on graft healing, we conducted a paraffin section test. At 2 DAG, a thin and deep-staining isolation layer was observed. The results show that low temperatures did not affect the formation of an isolated layer. With graft junction development, the isolation layer gradually disappeared. Callus tissue (CA) provides a pathway for communication between the scion and stock. The graft junction under a night temperature of 28 °C formed callus tissue at 3 DAG (Figure 1E), whereas callus formation under a night temperature of 18 °C was observed at 4 DAG (Figure 1B). Vascular connection between the grafted partners was a mark of grafting success. At 6 DAG (Figure 1F), the graft junction under a night temperature of 28 °C formed vascular bundles (VBs), and new vascular bundle formation under a night temperature of 18 °C occurred at 8 DAG (Figure 1C). The results suggest that the healing process of oriental melon scion grafted onto pumpkin rootstock was enhanced under a 28 °C night temperature compared to an 18 °C night temperature.

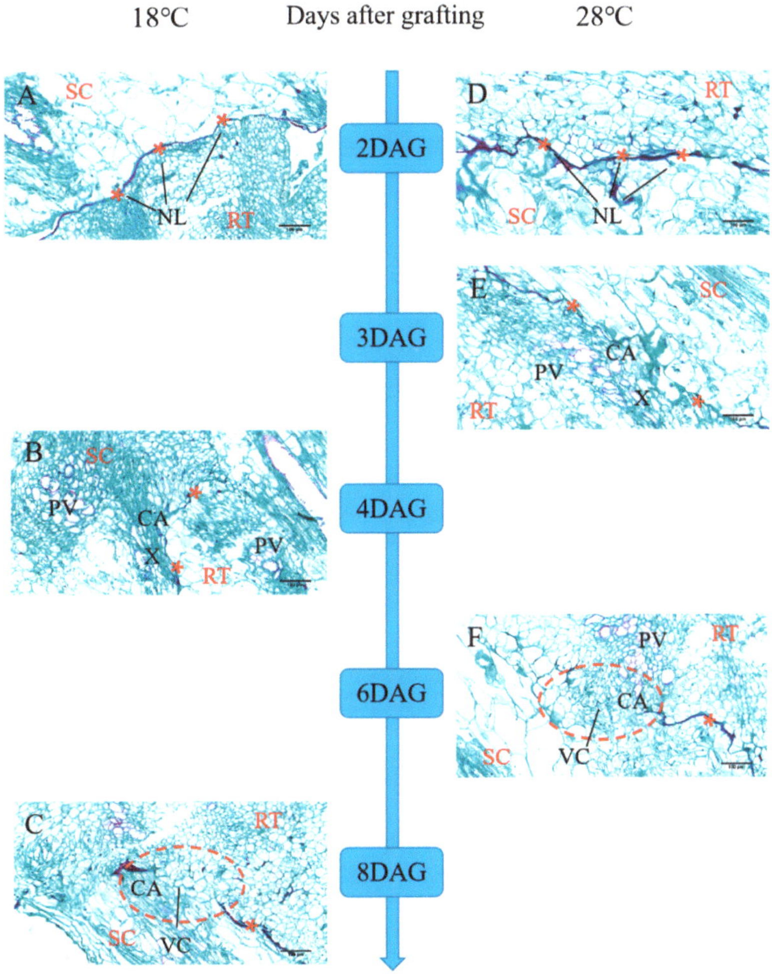

Figure 1. Histological changes of the graft junction using paraffin sectioning and microscopy methods during graft union development under 28 °C and 18 °C night temperatures. (**A**,**D**) Isolation layer stages; (**B**,**E**) the callus phase; (**C**,**F**) the vascular bundle reconnection phase; SC, scion; RT, rootstock; DAG, days after grafting; NL, necrotic layer; VC, vascular bundle reconnection; CA, callus; X, nascent xylem element; PV, original vascular bundle bridge; *, the graft interface; bars, 100 μm.

2.2. Reconnection of Vascular Bundles

The Esculin assay was used to monitor phloem connectivity. We applied Esculin solution to the cotyledons and examined the fluorescent signals on a daily basis. Upon comparing the epicotyl of the scion 1 cm above the graft junction with the hypocotyl of the rootstock 1 cm below the graft junction, few grafted individuals exhibited a fluorescence signal in the hypocotyl of the rootstock after the application of Esculin to the cotyledons at 2 DAG with a 28 °C night temperature treatment and 4 DAG with an 18 °C night temperature treatment, respectively. Nearly 90% of the individuals showed a fluorescence signal at 5 DAG under a night temperature of 28 °C and at 8 DAG under a night temperature of 18 °C (Figure 2A).

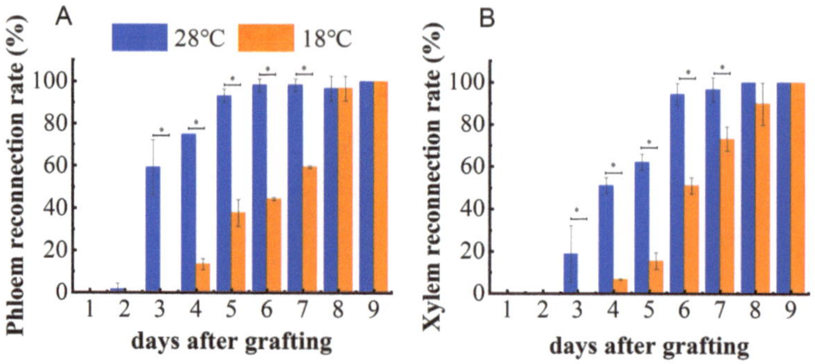

Figure 2. Phloem reconnection rate (**A**) and xylem reconnection rate (**B**) after grafting. Data fo each time point were collected from the three treatments with 20 seedlings per treatment and ar presented as the mean ± SE. A temperature of 28 °C means that the day and night temperatures wer 28 °C/28 °C. A temperature of 18 °C means that the day and night temperatures were 28 °C/18 °C Asterisks indicate significant differences using a *t*-test ($p \leq 0.05$).

Next, we soaked the rootstock in 0.1% (w/v) acid fuchsin solution and monitored th dye in the epicotyl of the scion 1 cm above the graft junction to assay xylem reconnectio Approximately 90% of scions exhibited acid fuchsin dye at 6 DAG under a night tempera ture of 28 °C and at 8 DAG under a night temperature of 18 °C (Figure 2B). Taken togethe the results show that phloem reconnection between the scion and rootstock under a nigh temperature of 28 °C was established two days earlier than under a night temperature o 18 °C, and xylem reconnection between the scion and rootstock under a night temperatur of 28 °C was established one day earlier than under a night temperature of 18 °C.

2.3. Variations of Sugars

To better understand how elevated temperatures promoted graft formation, we ana lyzed the content of sugars using an ultra-performance liquid chromatography–electrospra tandem mass spectrometry detection platform. The levels of 29 sugars in metabolic profile were analyzed throughout the graft union formation process. We compared the metaboli profiles of 28 °C and 18 °C at four stages to identify the DEMs during graft union formatio The DEMs were filtered according to an expression level $|\log_2$ (fold-change)$| > 1$ and *p*-value < 0.05 in each pairwise comparison. Nine sugars, including sucrose, trehalose maltose, D–ribono–1, 4–lactone, inositol, D–glucose, D–galactose, D–fructose, D–galactose and raffinose, were significantly enriched in graft union (Figure 3).

The content of sucrose, trehalose, raffinose, D–glucose, D–fructose, D–galactose, anc inositol showed an initial increase but then decreased. The content of sucrose, trehalose, anc raffinose reached a maximum on the second day after grafting, whereas under a night tem perature of 18 °C, they reached a maximum on the third day after grafting (Figure 3A–C Treatment with a 28 °C night temperature increased the content of sucrose, trehalose, anc raffinose significantly at 1 DAG and 2 DAG, but then the content of these sugars decrease significantly at 3 DAG and 5 DAG, compared with 18 °C night temperature treatment. O the second day after grafting, the content of D–glucose, D–fructose, D–galactose, and inosi tol in two treatments were highest (Figure 3D–G). Furthermore, the content of D–glucose D–fructose, D–galactose, and inositol under a night temperature of 28 °C was significantl increased, with levels 64.76%, 52.50%, 62.97%, and 26.33% higher than those under a nigh temperature of 18 °C, respectively.

Grafting induced D–ribono–1, 4–lactone synthesis-related metabolism, leading to a elevated content of D–ribono–1, 4–lactone. However, D–ribono–1, 4–lactone levels under a night temperature of 18 °C were significantly higher than those under a night temperatur of 28 °C (Figure 3H). The content of maltose under a night temperature of 28 °C wa stable in the early stages (1–3 days after grafting) and significantly increased 5 days after grafting (Figure 3I).

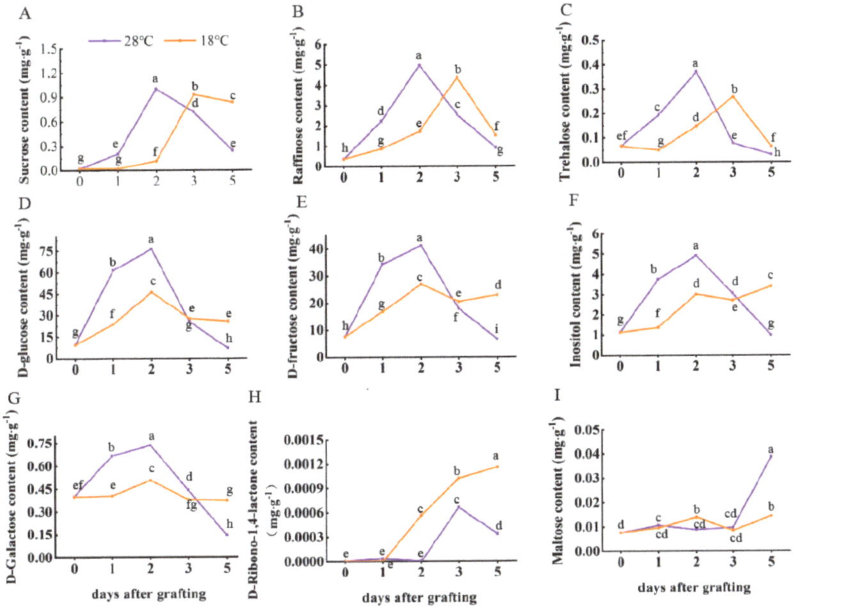

Figure 3. Nine significantly enriched sugar levels. (**A**) Sucrose content; (**B**) raffinose content; (**C**) trehalose content; (**D**) D–glucose content; (**E**) D–fructose content; (**F**) inositol content; (**G**) D–galactose content; (**H**) D–ribono–1, 4–lactone content; and (**I**) maltose content. The different small letters indicate significant differences at the $p < 0.05$ level.

2.4. Exogenous Sucrose, Glucose, or Fructose Treatment under 18 °C Night Temperature

To investigate the effects of the level and type of sugars on graft union formation, we sprayed a range of concentrations of exogenous sucrose, glucose, or fructose solution (0, 0.5, 1, and 2%) on the seedling after grafting. The results show that exogenous sucrose application could not promote vascular reconnection. Phloem reconnection was not significantly different under exogenous sucrose treatments (Figure 4A), and the xylem reconnection rate decreased by 466.67%, 277.78%, and 277.78% at 4 DAG under 0.5%, 1%, and 2% exogenous sucrose treatments compared with CK (Figure 4D). During the entire healing period after grafting, the xylem reconnection rate was not significantly different under exogenous sucrose treatments.

Exogenous glucose application promoted vascular reconnection (Figure 4B). The phloem reconnection rate increased by 94.12%, 94.12%, and 58.82% at 4 DAG under 0.5%, 1%, and 2% exogenous glucose treatments compared with CK. The xylem reconnection rate increased by 117.24% at 4 DAG under 0.5% exogenous glucose treatment compared with CK (Figure 4E). However, xylem reconnection was not significantly different under 1% and 2% exogenous glucose treatments.

Exogenous fructose application promoted phloem reconnection. The phloem reconnection rate increased by 26.92%, 38.46%, and 44.23% at 5 DAG under 0.5%, 1%, and 2% exogenous fructose treatments compared with CK (Figure 4C). The xylem reconnection rate was not significantly different under exogenous sucrose treatments (Figure 4F).

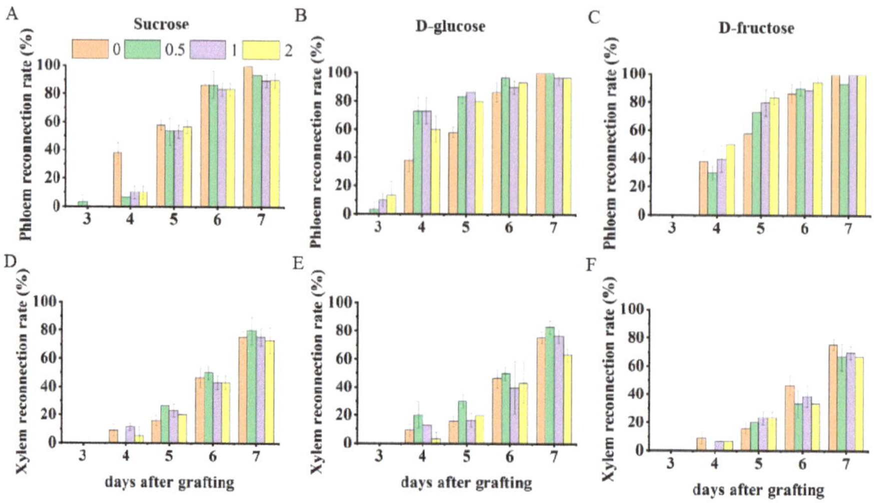

Figure 4. The phloem and xylem reconnection rate of grafted seedlings treated with exogenous sucrose, glucose, and fructose. (**A**,**D**) Exogenous sucrose treatment; (**B**,**E**) exogenous glucose treatment; and (**C**,**F**) exogenous fructose treatment.

3. Discussion

Temperature is the key factor for graft union formation. Several reports have shown that the optimal temperature range for healing grafted melons is 28–30 °C [17,18]. Previous studies have shown that when the night healing temperature decreases to 12 °C, there is a wider gully between the rootstock and scion cut surface and the differentiation of vascular tissue is delayed, leading to unsuccessful graft union formation [11]. Furthermore, the effects of different night temperatures on phloem and xylem reconnection were tested. In preliminary tests using 18 °C, 23 °C, and 28 °C, the differences between the phloem and xylem reconnection rates between 18 °C and 23 °C were not significant, but they were significant between 18 °C and 28 °C. Therefore, in this study, 28 °C was set as the high temperature and 18 °C was set as the low temperature to explore the effects of night temperature on the anatomy and sugar contents during the graft healing of melon–pumpkin heterografts.

Sucrose may improve callus formation and the connectivity of vascular bundles at the graft interface. Melnyk et al. [19] found that a level of 0.5% exogenous sucrose enhances the graft survival rate of Arabidopsis. Dabirian and Miles [20] found that drench applications of 1% and 2% sucrose solution to rootstock seedlings before grafting increases the carbohydrate level in the hypocotyl and increases grafting success when both cotyledons are removed from the rootstock before grafting. Miao et al. [14] found that phloem reconnection of cucumber/pumpkin grafts was not significantly different under exogenous glucose treatments, and xylem reconnection was achieved 1 day earlier after grafting with 0.5% exogenous glucose. Differential sugar responses at the graft junction might be important for vascular reconnection. Therefore, to test the effect of different sugars on graft healing, the grafted melon/pumpkin was sprayed with different concentrations of glucose (0.5, 1, and 2%) (w/v), fructose (0.5, 1, and 2%) (w/v), and sucrose (0.5, 1, and 2%) (w/v) when grafting.

3.1. *Elevated Temperature Improves the Process of Graft Union Formation*

Graft healing is a complex process that is affected by environmental factors, with temperature being one of the most important influencing factors. Previous work has described the stages of graft union formation, including the production of a necrotic layer, the proliferation of callus cells at the graft interface, and vascular redifferentiation across the graft interface [7,19,21]. The connection of the graft union between rootstock and scion is influenced by plant growth conditions, age, species, and so on. It takes an average of 5 to 14 d for the graft union of herbaceous plants to develop vascular connectivity between rootstock and scion [22,23].

Vascular tissue reconnection is considered an important indicator of grafting success. A robust vascular connection between the scion and the rootstock determines the physiological functionality of the plant, influencing vital processes such as water and nutrient uptake and translocation, which in turn regulate nutrition, organ growth, photosynthesis, and transpiration [24]. Oriental melon grafted onto squash showed union bridges between the rootstock and scion 9 days after grafting under 26 ± 3 °C temperature [25], and watermelon grafted onto squash showed that vascular connections appeared 5 days after grafting under a night temperature of 18 °C [11]. In this study, functional phloem connection occurred two days earlier than xylem connection when treated with a night temperature of 28 °C. This is consistent with what has been reported for grafted cucumber [14]. However, functional phloem and xylem connection occurred on the same day under a night temperature of 18 °C. These results imply that elevated temperatures improve the process of graft union formation.

3.2. The Levels of Nine Sugars Affect Graft Union Formation

A complex network of physiological responses takes place during the formation of the graft union and collectively determine the timeline and outcome of the procedure. Grafting causes mechanical damage to the plant, which sets off responses to counteract the damage, including the inhibition of cell elongation, stomatal closure, and reductions in transpiration, stomatal conductance, and CO_2 assimilation. Furthermore, there is an accumulation of reactive oxygen species (ROS) [26–28]. These ROS have the potential to cause oxidative damage. Accumulated soluble sugars can protect themselves against potential oxidative damage, including cellular membrane collapse and the breakdown of cellular compartments [29].

Achieving homeostasis allows the plant to rebuild tissues and forge a union between the two components of the graft. Tissue adhesion begins with the accumulation of polysaccharides, such as cellulose and pectin [30,31]. Carbon skeletons are required in the damaged area for the synthesis of new molecules [32]. Simultaneously, callus-like masses of undifferentiated cells (known as callus tissue) proliferate adjacent to the existing vascular tissue [33]. Furthermore, cell proliferation is an energy-consuming process. Sugars, as an energy source, provide ATP through glycolysis pathways and cellular respiration, and eventually activate cell metabolism. Sugars are actively involved in cell division and cell expansion [14,34]. Additionally, sugars, as signaling molecules, may regulate gene expression and a variety of metabolic processes through hexokinase-dependent and independent pathways. Pu et al. [35] found that the starch content in the rootstock cotyledons of cucumber grafted onto pumpkin seedlings decreased during 0–3 days after grafting. Amri et al. [36] discovered a significant increase in soluble sugar and starch content at the rootstock and scion union in peach/plum grafting. In our data sets, nine sugars were significantly increased. However, the varying trends observed in the levels of the nine sugars may have resulted in the abnormal formation of vascular tissue.

The accumulation of sugars in sink tissues depends on the actions of several genes related to biosynthesis, metabolism, and transportation. In *Cucurbitaceae*, raffinose and stachyose are the predominant carbohydrates transported from leaves to sink tissues [37]. In sink tissues, raffinose family oligosaccharides (RFOs) are unloaded from the phloem, and alkaline α-galactosidases hydrolyze the RFOs to sucrose and galactose. When a plant suffers stress, sucrose will be used for long-distance transport [38]. In this study, the change in sucrose content was similar to that of raffinose, but different from that of D–galactose. So, sucrose and raffinose were the main carbohydrates transported from leaves to the graft union. The exogenous application of sucrose did not promote vascular reconnection.

Sucrose can be converted to its hexose monomers (glucose and fructose) by invertases or converted to fructose and uridine diphosphate-glucose (UDPG) by sucrose synthase [39].

The metabolic processes of sucrose are mainly regulated by sucrose–phosphate synthas (SPS), sucrose–phosphate phosphatase (SPP), and sucrose synthase (SUS). The content D–glucose, D–fructose, D–galactose, and inositol in two treatments were highest 2 day after grafting. However, the contents of D–glucose, D–fructose, D–galactose, and inosit under a night temperature of 28 °C were significantly higher than those under a nigh temperature of 18 °C. Exogenous glucose application promoted vascular reconnectio and exogenous fructose application promoted phloem reconnection. Therefore, high nigh temperatures accelerate graft union formation by promoting the accumulation of glucos and fructose. However, further study is required to understand how glucose and fructos regulate graft union formation.

4. Materials and Methods

4.1. Plant Materials

Scion: melon (*Cucumis melo* L.), Meinong Cultivar, was obtained from Hubei Xueyi Agricultural Technology Co., Ltd., Jingzhou, China.

Rootstock: pumpkin (*Cucurbita moschata* Dutch.), Zhenzhuang Cultivar, was obtaine from Jingyan Yinong Seed Sci-Tech Co., Ltd., Shouguang, China.

All the experiments were carried out in an artificial chamber in 2022 at the Wuha Agricultural Academy, Central China (30°27′ N, 114°20′ E, and at an altitude of 22 m abov sea level). The scion and rootstock seeds were sown into 98- and 72-cell trays, respectivel with one seed in one cell filled with mixed seedling substrate (peat moss and pearlit at a volume ratio of 3:1). The melon seeds were sown 5 days before the pumpkin seed and were placed in a germination chamber at 30 °C for 2 days. After germination, th melon and pumpkin seedlings were transferred to the artificial chamber. The plants wer cultivated with a day/night (12 h/12 h) cycle at 28 °C/18 °C and 60–80% relative humidit The plants were fertilized with a water-soluble fertilizer (Product Model: 20-10-20 + TE 1000 times liquid, Shanghai Yongtong chemical Co., Ltd., Shanghai, China).

4.2. Grafting and Temperature Treatment

The melon scions were grafted onto pumpkin rootstock using the root-pruning splic grafting technique at the first-true-leaf development stage [2]. After grafting, the grafte melon seedlings were divided into two groups and were placed in a hydroponic bo containing 1/2 Hoagland's nutrient solution (12 cm × 12 cm) for continuous temperatur treatment. One group was placed in an artificial chamber with a temperature of 28 °C day/28 °C night (28 °C), and another group was placed in an artificial chamber with a temperature of 28 °C day/18 °C night (18 °C). The other conditions included a photoperio of 12 h day/12 h night, a light intensity of 50 μM·m^{-2}·s^{-1}, and a constant humidity o 95–100%. Three separate biological experiments were conducted under the same condition to replicate the results.

4.3. Exogenous Glucose, Fructose, and Sucrose Treatments

To test the effects of different sugars on graft healing, the grafted melon/pumpkin wa sprayed with different concentrations of glucose (0.5, 1, and 2%) (*w*/*v*), fructose (0.5, 1, an 2%) (*w*/*v*), and sucrose (0.5, 1, and 2%) (*w*/*v*) when grafting. Distilled water was spraye as the control. Three hundred plants were used for each treatment. Phloem and xylem connectivity was observed 3, 4, 5, 6, and 7 d after grafting. Twenty plant replicates wer performed for every treatment. Three separate biological experiments were conducte under the same conditions to replicate the results.

4.4. Paraffin Sectioning and Microscopy

A sample of 0.3 cm stems above and below the graft junction was collected at 1–9 DAG The collected samples were fixed, softened, dehydrated, infiltrated, and embedded in paraffin, as described by Ribeiro et al. [40]. The samples were sectioned to 10 μm verticall using a rotary microtome (RM2016; Leica; Shanghai, China), dewaxed, rehydrated, cleaned stained with Fast Green, counterstained with safranin, and then fixed with neutral balata. The sections were examined using a light microscope (NIKON DS-U3; Nikon Corp., Tokyo, Japan), and representative sections were photographed.

4.5. Assays of Phloem and Xylem Connectivity

Phloem and xylem connectivity were measured by Esculin and acid fuchsin movement across the graft union, respectively [41]. To assay phloem connections, 0.4 g of Esculin was dissolved in 20 mL of 60% Acetonitrile. The cotyledon center of the scion was gently scraped with a sharp single-edge razor to create a small opening for the Esculin to enter. Next, a 30 μL Esculin sample was added to each cotyledon, and fluorescence in the rootstock hypocotyls was measured after 2 h of incubation. For xylem connection, the plant roots were incubated in a solution of 5 mg/mL of acid fuchsin, and the melon hypocotyls were examined after 2 h.

The phloem and xylem reconnection rates were investigated using the following formulae:

$$\text{Phloem reconnection rate (\%)} = (\text{phloem reconnected grafts/total test grafts}) \times 100\%$$

$$\text{Xylem reconnection rate (\%)} = (\text{xylem reconnected grafts/total test grafts}) \times 100\%$$

4.6. Determination of Sugar Contents

The samples were harvested from the grafting union at 0, 1, 2, 3, and 5 DAG. The sugars were extracted and assayed according to previous reports [42]. All the metabolites were detected using MetWare (http://www.metware.cn/, accessed on 25 September 2022) on the AB Sciex QTRAP 6500 LC-MS/MS platform. After the sample was thawed and smashed, 0.05 g of the sample was mixed with 500 μL of 70% methanol/water. The sample was vortexed for 3 min under 2500 r/min and centrifuged at 12,000 r/min for 10 min at 4 °C. A total of 300 μL of supernatant was transferred into a new centrifuge tube and placed in a −20 °C refrigerator for 30 min. Then, the supernatant was centrifuged again at 12,000 r/min for 10 min at 4 °C. After centrifugation, 200 μL of supernatant was transferred through a Protein Precipitation Plate for further LC-MS analysis. The sample extracts were analyzed using an LC-ESI-MS/MS system (Waters ACQUITY H-Class, https://www.waters.com/nextgen/us/en.html, accessed on 18 December 2022; MS, QTRAP® 6500+ System, https://sciex.com/, accessed on 20 December 2022). Three separate biological experiments were conducted under the same conditions to replicate the results.

4.7. Statistical Analysis

The data are presented as the mean ± standard deviation of three replicate samples. Significantly regulated metabolites between the groups were determined using VIP and absolute Log_2FC (fold change). The statistical evaluations were performed via one-way analysis of variance, followed by individual comparisons with *t*-tests ($p < 0.05$) using SAS 9.0.3 software.

5. Conclusions

The graft healing process takes longer and involves higher costs in early spring, when plants often suffer from low temperature exposure. Temperature is undoubtedly the key factor affecting graft union formation. Elevated temperatures improve the process of graft union formation by increasing the content of sucrose, trehalose, maltose, D–ribono–1, 4–lactone, inositol, D–glucose, D–galactose, D–fructose, D–galactose, and raffinose. Moreover, exogenous glucose and fructose application promotes vascular reconnection. However, exogenous sucrose application does not promote vascular reconnection. These results provide information to develop strategies for improving grafting efficiency under low temperatures.

Author Contributions: Conceptualization, H.L. and M.Z.; methodology, H.L.; validation, J.L., X.S. and J.Z.; formal analysis, J.L.; investigation, H.L., J.Z., D.W. and M.G.; data curation, H.L., M.Z. and X.S.; writing—original draft preparation, H.L. and J.L.; writing—review and editing, H.L. and M.Z.; supervision, X.S.; funding acquisition, X.S. All authors have read and agreed to the published version of the manuscript.

Funding: This research was supported by the Key R&D projects in Hubei Province (2023BBB033) and the Knowledge Innovation Program of Wuhan-Shuguang Project (2022020801020418).

Institutional Review Board Statement: Not applicable.

Informed Consent Statement: Not applicable.

Data Availability Statement: The data presented in this study are available upon request from the first author at lianghuanconf@126.com.

Conflicts of Interest: The authors declare no conflicts of interest.

References

1. Lee, J.M.; Kubota, C.; Tsao, S.J.; Bie, Z.L.; Hoyos Echevarria, P.; Morra, L.; Oda, M. Current status of vegetable grafting: Diffusion, grafting techniques, automation. *Sci. Hortic.* **2010**, *127*, 93–105. [CrossRef]
2. Davis, A.R.; Perkins-Veazie, P.; Sakata, Y.; López-Galarza, S.; Maroto, J.V.; Lee, S.-G.; Huh, Y.-C.; Sun, Z.; Miguel, A.; King, S.; et al. Cucurbit Grafting. *Crit. Rev. Plant Sci.* **2008**, *27*, 50–74. [CrossRef]
3. Nawaz, M.A.; Wang, L.; Jiao, Y.; Chen, C.; Zhao, L.; Mei, M.; Yu, Y.; Bie, Z.; Huang, Y. Pumpkin rootstock improves nitrogen use efficiency of watermelon scion by enhancing nutrient uptake, cytokinin content, and expression of nitrate reductase genes. *Plant Growth Regul.* **2017**, *82*, 233–246. [CrossRef]
4. Yan, G.; Feng, M.; Lin, W.; Huang, Y.; Tong, R.; Cheng, Y. Review and Prospect for Vegetable Grafting Robot and Relevant Key Technologies. *Agriculture* **2022**, *12*, 1578. [CrossRef]
5. Loupit, G.; Brocard, L.; Ollat, N.; Cookson, S.J. Grafting in plants: Recent discoveries and new applications. *J. Exp. Bot.* **2023**, *74*, 2433–2447. [CrossRef]
6. Liu, Y.; Liu, H.; Zhang, T.; Liu, J.; Sun, X.; Sun, X.; Wang, W.; Zheng, C. Interactions between rootstock and scion during grafting and their molecular regulation mechanism. *Sci. Hortic.* **2023**, *308*, 111554. [CrossRef]
7. Feng, M.; Augstein, F.; Kareem, A.; Melnyk, C.W. Plant grafting: Molecular mechanisms and applications. *Mol. Plant* **2024**, *17*, 75–91. [CrossRef]
8. Avanzato, D.; Tamponi, G. The effect of heating of walnut graft unions on grafting success. *Acta Hortic.* **1988**, *227*, 79–83. [CrossRef]
9. Serivichyaswat, P.T.; Bartusch, K.; Leso, M.; Musseau, C.; Iwase, A.; Chen, Y.; Sugimoto, K.; Quint, M.; Melnyk, C.W. High temperature perception in leaves promotes vascular regeneration and graft formation in distant tissues. *Development* **2022**, *149*, dev200079. [CrossRef]
10. Shibuya, T.; Nakashima, H.; Shimizu-Maruo, K.; Kawara, T. Improvement of Graft Development in Tomato and Eggplant Grafted Cuttings by Supplying Warmed Water to Graft Union during Low-air-temperature Storage. *J. Jpn. Soc. Hortic. Sci.* **2007**, *76*, 217–223. [CrossRef]
11. Yang, X.; Hu, X.; Zhang, M.; Xu, J.; Ren, R.; Liu, G.; Yao, X.; Chen, X. Effect of low night temperature on graft union formation in watermelon grafted onto bottle gourd rootstock. *Sci. Hortic.* **2016**, *212*, 29–34. [CrossRef]
12. Devi, P.; Lukas, S.; Miles, C. Advances in Watermelon Grafting to Increase Efficiency and Automation. *Horticulturae* **2020**, *6*, 88. [CrossRef]
13. Nie, W.; Wen, D. Study on the Applications and Regulatory Mechanisms of Grafting on Vegetables. *Plants* **2023**, *12*, 2822. [CrossRef]
14. Miao, L.; Li, Q.; Sun, T.-S.; Chai, S.; Wang, C.; Bai, L.; Sun, M.; Li, Y.; Qin, X.; Zhang, Z.; et al. Sugars promote graft union development in the heterograft of cucumber onto pumpkin. *Hortic. Res.* **2021**, *8*, 146. [CrossRef]
15. Almansa, E.M.; Chica, R.M.; Lao, M.T. Influence of the quality of artificial light on grafting tomato. *Aust. J. Crop Sci.* **2018**, *12*, 318–325. [CrossRef]
16. Marsch-Martínez, N.; Franken, J.; Gonzalez-Aguilera, K.L.; Folter, S.; Angenent, G.; Alvarez-Buylla, E.R. An efficient flat-surface collar-free grafting method for *Arabidopsis thaliana* seedlings. *Plant Methods* **2013**, *9*, 14. [CrossRef]
17. Hassell, R.; Memmott, F.; Liere, D. Grafting methods for watermelon production. *HortScience* **2008**, *43*, 1677–1679. [CrossRef]
18. Johnson, S.J.; Miles, C.A. Effect of healing chamber design on the survival of grafted eggplant, tomato, and watermelon. *HortTechnology* **2011**, *21*, 752–758. [CrossRef]
19. Melnyk, C.W. Plant grafting: Insights into tissue regeneration. *Regeneration* **2016**, *4*, 3–14. [CrossRef]
20. Dabirian, S.; Miles, C.A. Increasing Survival of Splice-grafted Watermelon Seedlings Using a Sucrose Application. *HortScience* **2017**, *52*, 579–583. [CrossRef]

1. Rasool, A.; Mansoor, S.; Bhat, K.M.; Hassan, G.I.; Baba, T.R.; Alyemeni, M.N.; Alsahli, A.A.; El-Serehy, H.A.; Paray, B.A.; Ahmad, P. Mechanisms Underlying Graft Union Formation and Rootstock Scion Interaction in Horticultural Plants. *Front. Plant Sci.* **2020**, *11*, 590847. [CrossRef]
2. Fan, J.; Yang, R.; Li, X.; Zhao, W.; Zhao, F.; Wang, S. The processes of graft union formation in tomato. *Hortic. Environ. Biotechnol.* **2015**, *56*, 569–574. [CrossRef]
3. Zhu, Y.; Hu, S.; Min, J.; Zhao, Y.; Yu, H.; Irfan, M.; Xu, C. Transcriptomic analysis provides aninsight into the function of CmGH9B3, a key gene of β-1,4-glucanase, during the graft union healing of oriental melon scion grafted onto squash rootstock. *Biotechnol. J.* **2024**, *19*, e2400006. [CrossRef] [PubMed]
4. Martínez, M.; Alcaraz, C.; Muries, B.; Mota, C.; Carvajal, M. Physiological aspects of rootstock–scion interactions. *Sci. Hortic.* **2010**, *127*, 112–118. [CrossRef]
5. Xu, C.; Zhang, Y.; Zhao, M.; Liu, Y.; Xu, X.; Li, T. Transcriptomic analysis of melon/squash graft junction reveals molecular mechanisms potentially underlying the graft union development. *PeerJ.* **2021**, *9*, e12569. [CrossRef]
6. Fernandez, N. Graft Union Formation in Tomato Plants: Peroxidase and Catalase Involvement. *Ann. Bot.* **2004**, *91*, 53–60. [CrossRef]
7. Carmach, C.; Castro, M.; Peñaloza, P.; Guzmán, L.; Marchant, M.J.; Valdebenito, S.; Kopaitic, I. Positive Effect of Green Photo-Selective Filter on Graft Union Formation in Tomatoes. *Plants* **2023**, *12*, 3402. [CrossRef]
8. Savatin, D.V.; Gramegna, G.; Modesti, V.; Cervone, F. Wounding in the plant tissue: The defense of a dangerous passage. *Front. Plant Sci.* **2014**, *5*, 470. [CrossRef]
9. Sami, F.; Yusuf, M.; Faizan, M.; Faraz, A.; Havat, S. Role of sugars under abiotic stress. *Plant Physiol. Biochem.* **2016**, *109*, 54–61. [CrossRef]
10. Frey, C.; Álvarez, R.; Encina, A.; Acebes, J. Tomato graft union failure is associated with alterations in tissue development and the onset of cell wall defense responses. *Agronomy* **2021**, *11*, 1197. [CrossRef]
11. Du, J.; Anderson, C.T.; Xiao, C. Dynamics of pectic homogalacturonan in cellular morphogenesis and adhesion, wall integrity sensing and plant development. *Nat. Plants* **2022**, *8*, 332–340. [CrossRef]
12. Morkunas, I.; Borek, S.; Formela, M.; Ratajczak, L. Plant responses to sugar starvation. *Carbohydr. Compr. Stud. Glycobiol. Glycotechnol.* **2012**, *19*, 409–438.
13. Frey, C.; Acebes, J.L.; Encina, A.; Álvarez, R. Histological changes associated with the graft union development in tomato. *Plants* **2020**, *9*, 1479. [CrossRef]
14. Wang, L.; Ruan, Y.L. Regulation of cell division and expansion by sugar and auxin signaling. *Front. Plant Sci.* **2013**, *4*, 163. [CrossRef] [PubMed]
15. Pu, D.; Zhang, J.; Lei, L.; Shang, Q.; Dong, C. Dynamics of Starch Metabolism in the Rootstock Cotyledon of Cucumber/Pumpkin Grafted Seedlings. *Acta Bot. Boreali-Occident. Sin.* **2022**, *11*, 1892–1901.
16. Amri, R.; Forcada, C.F.I.; Gimenez, R.; Pina, A.; Moreno, M. Biochemical characterization and differential expression of pal genes associated with "translocated" peach/plum graft-incompatibility. *Front. Plant Sci.* **2021**, *12*, 622578. [CrossRef]
17. Yan, S.; Liu, Q.; Li, W.; Yan, J.; Fernie, A.R. Raffinose Family Oligosaccharides: Crucial Regulators of Plant Development and Stress Responses. *Crit. Rev. Plant Sci.* **2022**, *41*, 286–303. [CrossRef]
18. Hao, J.H.; Li, T.L.; Sun, L.P.; Zhao, B.; Meng, S.D. Effects of Night Low Temperature on Carbohydrate Content in Different Organs along Assimilate Transport Path of Melon. *Acta Bot. Boreal.* **2009**, *29*, 0085–0092.
19. Ruan, Y.L. Signaling role of sucrose metabolism in development. *Mol. Plant* **2012**, *5*, 763–765. [CrossRef]
20. Ribeiro, L.M.; Nery, L.A.; Vieira, L.M.; Mercadante-Simões, M.O. Histological study of micrografting in passionfruit. *Plant Cell Tissue Organ Cult. (PCTOC)* **2015**, *123*, 173–181. [CrossRef]
21. Xu, J.N.; Wei, X.Y.; Xiong, M.; Zhang, T.; Liu, C.J.; Bie, Z.L.; Huang, Y. A method for simultaneously monitoring phloem and xylem reconnections in grafted watermelon seedlings. *Sci. Hortic.* **2022**, *299*, 111058. [CrossRef]
22. Yuan, H.; Zeng, X.; Yang, Q.; Xu, Q.; Wang, Y.; Jabu, D.; Sang, Z.; Tashi, N. Gene coexpression network analysis combined with metabonomics reveals the resistance responses to powdery mildew in Tibetan hulless barley. *Sci. Rep.* **2018**, *8*, 14928. [CrossRef] [PubMed]

plants

MDPI

Article

Growth and Physiological Characteristics of Strawberry Plants Cultivated under Greenhouse-Integrated Semi-Transparent Photovoltaics

Theodoros Petrakis [1], Paraskevi Ioannou [2], Foteini Kitsiou [3], Angeliki Kavga [1,*], George Grammatikopoulos [3] and Nikos Karamanos [2,4]

1 Department of Agriculture, University of Patras, 26504 Patras, Greece; tpetrakis@ac.upatras.gr
2 Biochemistry, Biochemical Analysis & Matrix Pathobiology Research Group, Laboratory of Biochemistry, Department of Chemistry, University of Patras, 26504 Patras, Greece; ioannouevi@icloud.com (P.I.); n.k.karamanos@upatras.gr (N.K.)
3 Laboratory of Plant Physiology, Department of Biology, University of Patras, 26504 Patras, Greece; kitsioufot@gmail.com (F.K.); grammati@upatras.gr (G.G.)
4 Foundation for Research and Technology-Hellas (FORTH)/Institute of Chemical Engineering Sciences (ICE-HT), 26504 Patras, Greece
* Correspondence: akavga@upatras.gr

Abstract: The integration of semi-transparent photovoltaics into the roof of greenhouses is an emerging technique used in recent years, due to the simultaneous energy and food production from the same piece of land. Although shading in many cases is a solution to maintain the desired microclimate, in the case of photovoltaic installations, the permanent shading of the crop is a challenge, due to the importance of light to the growth, morphogenesis, and other critical physiological processes. In this study, the effect of shade from semi-transparent photovoltaics on a strawberry crop (*Fragaria ananassa* Duch.) was examined, in terms of growth and quality (phenolic and flavonoid concentration of fruits). According to the results, in non-shaded plants, there was a trend of larger plants, but without a significant change in leaf number, while the total number of flowers was slightly higher at the end of the cultivation period. Moreover, it was found that the percentage change between the number of ripe fruits was smaller than that of the corresponding change in fruit weight, implying the increased size of the fruits in non-shaded plants. Finally, regarding the antioxidant capacity, it was clearly demonstrated that shading increased the total phenolic content, as well as the free radical-scavenging activity of the harvested fruits. Although the shading from the semi-transparent photovoltaics did not assist the production of large fruits, it did not affect their number and increased some of their quality characteristics. In addition, the advantageous impact of the semi-transparent photovoltaics in the energy part must not be neglected.

Keywords: *Fragaria x ananassa* Duch.; shading; agrivoltaics; crop performance; total phenolic content; antioxidant activity

check for updates

Citation: Petrakis, T.; Ioannou, P.; Kitsiou, F.; Kavga, A.; Grammatikopoulos, G.; Karamanos, N. Growth and Physiological Characteristics of Strawberry Plants Cultivated under Greenhouse-Integrated Semi-Transparent Photovoltaics. *Plants* **2024**, *13*, 768. https://doi.org/10.3390/plants13060768

Academic Editors: Francesco Serio, Weijie Jiang and Wenna Zhang

Received: 25 January 2024
Revised: 5 March 2024
Accepted: 6 March 2024
Published: 8 March 2024

1. Introduction

Strawberry plants generally thrive in well-drained soil and temperate climates. The Mediterranean climate is well-suited for strawberry cultivation due to its mild, wet winters and hot, dry summers. Strawberry plants are often grown in greenhouses either for protection from adverse environmental conditions or, more commonly, to extend the growing season [1]. Light and temperature inside the greenhouse are the most influencing factors that affect the growth of strawberry plants [2]. The optimum light conditions vary according to the time of year and the growth stage of the plant. Light intensity is important for the growth, morphogenesis, and other physiological processes of plants by affecting the rate of photosynthesis, the number of leaves, the stem length, the branching, the commencement of flowering, and the fruit development and ripening [3–5]. In particular, low light intensity can reduce the rate of photosynthesis in plants, resulting in limited growth and lower yields [6]. The effects of reduced light intensity on the quality of harvested fruits are also important because strawberries have a variety of secondary metabolites, like phenolics and flavonoids. These metabolites serve as plant tissue defense against biotic and abiotic stresses, preventing cellular damage caused by reactive oxygen species [7].

Fruit berries, especially strawberries, are well-known sources of bioactive compounds (BACs). The major components of BACs in strawberries are phenolic compounds, which include phenolic acids (such as hydroxybenzoic and hydroxycinnamic acids), flavonoids (anthocyanins), flavonols (quercetin, kaempferol, myricetin), and flavanols (catechins and epicatechin) [8,9]. These components along with ascorbic acid are the most abundant antioxidants present in strawberries. Antioxidants work by donating hydrogen to free radicals, scavenging them, and generating more stable antioxidant radicals [10].

Shading in greenhouses protects against excess solar radiation during hot periods while also posing challenges during periods of limited sunshine and low temperatures [11]. The fluctuations of environmental conditions like these are relevant in Mediterranean regions. In the summer and in regions such as Greece, the temperature rises dramatically inside a greenhouse, with its values approaching 60 °C for extended periods (often from May to October), causing severe crop issues. Shading prevents the overheating of plants and reduces evapotranspiration by reducing incident light intensity and, consequently, lowering the temperature inside the greenhouse. Yet, shading is generally considered a limiting factor for growth and yields [12,13].

Aside from conventional greenhouse shading solutions such as plastic nets [14–17] and screens [18,19], one approach that has seen significant expansion and application in recent years is the installation of photovoltaic modules on the greenhouse's roof. Solar radiation is the most important parameter in satisfying production performance because photosynthesis is a biological process that significantly depends on sunlight. Despite that, the strength of an energy production system using solar radiation, such as a photovoltaic system, is primarily determined by the strength of the radiation incident on it. The requirements of both the crop and the solar system for sunlight make greenhouses the perfect structure, as they are placed in areas with a lack of surrounding impediments. Simultaneously, significant challenges have been encountered because of the increased demand for accessible land, both spatially and economically. Thus, using the same piece of land for both food and energy production appears to be a perfect solution, and this combination is also known as an agrivoltaic system [11,20,21].

According to the research findings, the employment of shade methods provides several advantages for preserving the greenhouse's microclimate at desirable levels in warm and cold regions. Indicatively, in [22], it is stated that, on the one hand, shading can maintain the temperature of the greenhouse in a range of 5 to 10 °C lower than that of the outside air, while this reduction combined with the effect of a cooling system increases the relative humidity up to 20%. At the same time, solar radiation can be reduced by up to 50%. On the other hand, in cold climate regions, the application of shading materials can act as additional insulation, reducing heat leaks, with the temperature being maintained up to 5 °C higher than outside.

Regarding the use of semi-transparent photovoltaics in a greenhouse, in [23], it is stated that, by covering the roof of a polyethylene-covered greenhouse with 20% photovoltaics, the solar radiation can be reduced by 35–40% during clear days, while the temperature reduction with the use of photovoltaics was 1–3 °C. At the same time, it is reported that, with an annual energy production equal to 637 kW h, the payback period was 9 years.

Several studies on the impact of semi-transparent PV panels on plant development in greenhouse conditions have been conducted. According to the literature, shading impacts different plant species in different ways, with [24,25] focusing on tomato cultivation, [2,26] on strawberries, [27] on berries, and [28] on lettuce, while there are also numerous studies demonstrating different outcomes. More specifically, Ref. [29] investigated four shading levels, which were the result of four different rates of coverage, with percentages amounting to 0%, 15%, 30%, and 50%. The study was performed for tomato cultivation and for a greenhouse with its ridge pointed in the north–south direction. The results showed that the reduction of solar radiation, due to the increase in the degree of coverage, had negative effects, both on the quantity and the quality of the fruits.

This study aimed to investigate the effect of greenhouse-integrated semi-transparent photovoltaics' shading on the parameters reflecting the size of the plant, the number of leaves, the flowers, the ripened fruits, and their weight. Moreover, the effect of shading on the antioxidant capacity (total phenolic content (TPC) and free radical scavenging activity) of the harvested fruits was examined.

2. Materials and Methods

2.1. Experimental Setup

2.1.1. Greenhouse

The cultivation of the *Fragaria x ananassa* Duch. plants was conducted in the experimental greenhouse of the Plant Physiology Lab (Dept of Biology, University of Patras) located on the University campus (38°17′27.9″ N, 21°47′23.9″ E) in a distinct unit that measures 3.2 m in width and 16.5 m in length, from 15 December 2022 to 1 June 2023.

The greenhouse's dimensions (gutter height of 3.22 m, ridge height of 3.9 m) are close to those of a real-scale greenhouse, with its type (even span) and east–west orientation making it physically identical to a commercial greenhouse. The greenhouse is made up of four identical, distinct units. It is a high-tech greenhouse made of premium materials and its frame is constructed of steel pieces with varying cross-sections. The cover material is glass, which is resistant to chemicals, high winds, and pollution. The greenhouse has a natural ventilation system with openings on the north and south inclined planes of the roof, as well as the side portions.

The construction unit used for the cultivation of strawberry plants is part of the overall greenhouse and, more specifically, is located on its northern side, with the area of covered ground equal to approximately 50 m^2. In this unit, natural ventilation is based only on windows on the northern sloped plane of the roof, due to the installation of photovoltaics on the southern sloped plane.

2.1.2. Photovoltaic Modules

The photovoltaic modules are characterized as semi-transparent, with permeability to sunlight and especially the spectrum of photosynthetically active radiation (PAR) (400–700 nm) being particularly high, while at the same time, intense light scattering is observed on the surface that does not contain solar cells. This scattering has the advantage of avoiding strong shadows from any skeletal elements that may lie beneath them.

Each photovoltaic module has dimensions of 2.089 m long and 1.033 m wide, a thickness of 5.5 mm, and a weight of 28.6 kg. The solar cells lying between two sheets of glass and within an encapsulation layer of POE/EVA material are equal to 80. Their dimensions correspond to 166 mm long and 83 mm wide, making an active surface of approximately 1.1 m^2. Finally, their bifacial type increases energy production as it is reliant on light falling both on the top and bottom sides of each cell. The nominal maximum power of each photovoltaic module is equal to 250 Wp, while according to the dimensions of both the whole module and the solar cells, the shading formed corresponds to approximately 51%.

2.1.3. Installation

Both in the arable unit and the unit next to it, twelve semi-transparent photovoltaics have been integrated. Their installation was performed by removing the old glass cover and replacing it by adding an aluminum bracket on the perimeter. The inclination angle of the photovoltaics is equal to $\sim$24°, so the inclination angle of the roof, while setting the units into the south sloped level of the roof increases the PVs' energy production, particularly during the winter, when the Sun makes a smaller angle to the ground. Out of twelve solar panels, eight have been installed on the roof of the arable greenhouse unit (Figure 1a), while the rest have been installed on the roof of the adjacent unit (Figure 1b). Their long side follows the east–west ridge direction, with them covering approximately 67% and 34% of the south sloped roof, respectively.

(**a**) (**b**)

Figure 1. Photovoltaic modules integrated into the roof of the greenhouse: (**a**) Greenhouse arable construction unit. (**b**) Adjacent greenhouse unit.

2.2. *Algorithm for Calculating the Shade*

To determine the percentage of shading caused by the photovoltaics inside the greenhouse and on the ground, an algorithm based on the specific greenhouse was used, which enables the detection of shading caused by photovoltaics on the ground for each time of year and for a time step of 10 min. The algorithm is based on astronomical equations and theorems relating to the position of the Sun, and it presents the shade using vector analysis and geometry. The algorithm has been validated using radiation data collected from within the greenhouse, according to all of its geometric and geographical aspects [30]. The outputs of the algorithm are the shaded area of the greenhouse in m^2, the percentage of ground covered by shade (%), and two graphical representations of the greenhouse, PVs, and shade, one 3D and one 2D. Finally, by providing its coordinates, it is possible to obtain a result showing whether a surface point is inside or outside the shaded area.

In this study, just the percentage of the shaded region was used from the overall output of the algorithm, which was executed for two separate scenarios, one where the area of interest concerned the planting line A and one where the area of interest concerned the planting line B. It should be noted that, although the algorithm extracts accurate results based on the arrangement according to which the PVs are placed in the greenhouse, the PVs are considered opaque. To calculate the percentage of shading formed by the photovoltaics on each line, the theoretical shading caused by them must be included. This shading is a result of the relationship between the active surface (total surface of the solar cells) and the total surface of the photovoltaic module (Equation (1)).

$$\textit{theoretical shading perc.}\ (\%) = \left(\frac{N \cdot E_{solar\ cell}}{E_{PV\ module}}\right) \times 100 \quad (1)$$

where N is the number of solar cells in each PV module equal to 80, $E_{solar\ cell}$ is the surface of the solar cell equal to 0.013 78 m^2, and $E_{PV\ module}$ is the total surface of each PV module equal to 2.158 m^2. According to the above, the resulting theoretical shading is equal to 51.08%, while this factor is also presented in the technical documentation accompanying the photovoltaic units, representing the semi-transparent nature of the modules. Thus, b multiplying the algorithm's output by the theoretical percentage of shading created by th semi-transparent photovoltaics, we can compute the shading formed by the photovoltaic on the surface of each planting line.

The mean daily percentage of shaded surface based on the foregoing is provided i Figure 2, from the day the plants were planted until the last day of the experiment. I addition to the percentage of the shaded surface, the mean daily internal temperature provided in Figure 2, with a loss of data for the period 13 January 2023 to 13 February 202 due to temperature sensor failure.

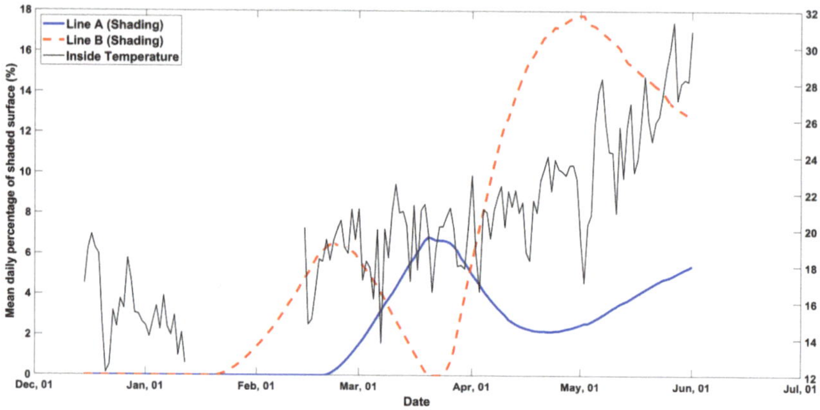

Figure 2. Mean daily percentage of shaded surface for planting lines A and B and internal temperatur

2.3. *Growth, Morphological, and Yield Measurements*

Frigo plants were planted in two single rows with an approximately 1.5 m distanc between rows. The length of each row was approximately 15 m; therefore, the spacin among the 20 plants used in each line was 75 cm. The soil was covered with black plasti mulch, and the plants were irrigated by a drip tube with a diameter of 20 mm under th mulch, delivering 2 L h^{-1} per dripper (Figure 3a,b). Each plant was equipped with on dripper, with the number of drippers corresponding to the number of plants. The irrigatio frequency was scheduled every three days for the period from 15 December 2022 to 1 Marc 2023 and daily application until the end of the experiment, to ensure that the soil was a the field capacity. Measurements were taken every 15 days beginning on 15 February 202 and the harvest on 15 March 2023 was extended by two days due to the unpredictable frui development. The measurements performed on the crop concern (a) the plant size throug three parameters: the plant size along the x-axis (along the planting line), the y-axis siz (perpendicular to the planting line), and the z-axis (plant height); (b) the number of leave (c) the number of flowers; (d) the number of red (ripened) fruits; and (e) the weight of th ripe strawberries.

2.4. *Sample Preparation and Extraction Methodology*

The strawberry samples utilized in this study were stored at −20 °C. For the extractio process, three strawberries from each line (denoted as A and B) were first allowed t thaw at room temperature (RT). Subsequently, each strawberry was weighed after carefu removal of the stem. Next, each strawberry was homogenized to achieve uniformity i mass, and the resulting product was reweighed. Then, 1 g of the homogenized strawberr was transferred to a falcon tube, and 12 mL of 70:30 [methanol (MeOH)–2x distilled (2d water] was added as the extraction solvent. The homogenate, along with the extractior solvent, was left for 1 h with frequent mild agitation in vortex to facilitate the extractior

Following this, centrifugation at 3000 rpm for 10 min at room temperature was performed to separate the supernatant, which was collected and used for chemical analyses.

(a) (b)

Figure 3. (**a**,**b**): Both subfigures represent strawberries' cultivation in the greenhouse.

2.5. TPC and DPPH Analyses

The TPC [31] in the strawberry extracts was obtained with Folin–Ciocalteu's reagent (Panreac, Barcelona, Spain). In this study, each sample (0.125 mL) was combined with 0.75 mL of 2d water and 0.125 mL of the Folin–Ciocalteu reagent. The resulting solution was allowed to incubate for 6 min. Subsequently, 2 mL of a 5% (w/v) sodium carbonate solution was added, and the mixture was left to incubate for an additional 90 min at room temperature in the absence of light. The absorbance of the samples was measured at 760 nm using a TECAN photometer. The obtained results are expressed as mg of gallic acid equivalent (GAE) per g of strawberry, representing the average of three measurements for each sample in triplicate.

The free-radical-scavenging activity of the extract was evaluated using the 1,1-diphenyl-2-picryl-hydrazyl (DPPH) method [32]. More precisely, a 0.2 mM DPPH solution (Cayman Chemical, Ann Arbor, MI, USA) was prepared fresh in MeOH. In this study, each sample (0.02 mL) was combined with 0.180 mL of the freshly prepared 0.2 mM DPPH solution, and it was allowed to incubate for 30 min at room temperature in the absence of light. The absorbance of the samples was measured at 517 nm using a TECAN photometer. The obtained results are expressed as mg of L-ascorbic acid equivalent (AAE) per g of strawberry, representing the average of three measurements for each sample in triplicate.

2.6. Gallic Acid and L-Ascorbic Acid Calibration Curves

To obtain a 1% solution of gallic acid (GA) (10 mg mL^{-1}), 0.15 g of GA (Sigma-Aldrich, St. Luis, MO, USA) was dissolved in 15 mL of 70:30 [methanol (MeOH)–2d water]. Subsequently, a standard gallic acid curve was established by diluting the standard solution of gallic acid to concentrations of 0.1, 0.5, and 1 mL in 70:30 [methanol (MeOH)–2d water]. The calibration curve was obtained using the TPC method, as described above. The equation obtained is represented by Equation (2) with a perfect linear regression of $R^2 = 1$.

$$y = 2.2657 \cdot x + 0.1342 \quad (2)$$

To obtain a 100 µg mL^{-1} solution of L-ascorbic acid (100 mg mL^{-1}), 0.015 g of L-ascorbic acid (Acros Organics, Fair Lawn, NJ, USA) was dissolved in 150 mL of 2d water. Afterward, a standard L-ascorbic acid curve was established by diluting the standard solution of L-ascorbic acid to concentrations of 5, 10, 20, 30, 50, 75, and 100 µg mL^{-1} in 2d water.

The calibration curve was obtained using the DPPH method, as described above. Th equation acquired is represented by Equation (3) with a linear regression of $R^2 = 0.9424$.

$$y = 0.6858 \cdot x + 4.21288 \quad (3$$

2.7. Statistical Analysis

The reported values for the growth, morphological, and yield measurements ar expressed as the mean ± the standard error (SE).

The reported values are expressed as the mean ± the standard deviation (SD) of th experiments in triplicate. Statistically significant differences were evaluated using Tukey' test to determine statistical differences between each data set of the two lines (A and B Differences were considered statistically significant at the level of $p \leq 0.05$, indicated by a asterisk (*). The statistical analysis and graphs were achieved using GraphPad Prism 8.0. (GraphPad Software, San Diego, CA, USA).

3. Results

3.1. Growth and Morphological Data

Using the three parameters relating to the size of each plant (x-, y-, and z-axis) an taking into account that the specified directions produce a Cartesian coordinate syster in space, it is feasible to define a vector, $\vec{v}_i$, in space and for each plant, i, of the forn (Equation (4)):

$$\vec{v}_i = (x_i \cdot \hat{x}, y_i \cdot \hat{y}, z_i \cdot \hat{z}) \quad (4$$

where x, y, and z are the plant size along the x-, y-, and z-axis, respectively, and $\hat{x}$, $\hat{y}$, and are the unit vectors in the directions along the planting line, perpendicular to the plantin line, and the plant height, respectively. The magnitude of $\vec{v}_i$ will be a number (in cm) tha according to Equation (5), will define the size of the plant by integrating the above thre parameters for the size of each plant.

$$|\vec{v}_i| = \sqrt{x_i^2 + y_i^2 + z_i^2} \quad (5$$

As shown in Figure 4, there was a trend of bigger plants in the line with the lowe overall shading (line A), though the differences between the two lines were not statisticall significant. The small, positive effect on the plant size for the plants in line A appearec 75 days from the beginning of the measurements (b.m.) when the cumulative shadin difference between the two lines was obvious.

The mean number of leaves was almost identical between the two planting row during the growth period (Figure 5a). The total number of leaves for each row at eac measurement date was not affected by the fluctuations of radiation during this perioc (Figure 5c), resulting in a similar cumulative number of leaves until the end of the ex periment (Figure 5e). The onset of flowering was simultaneous in the two rows, and th mean number of flowers per plant was similar between the planting rows during th experimental period (Figure 5b). As shown in Figure 5d, ascending phases of radiatior positively affected the production of inflorescences in the corresponding planting row Consequently, the final total number of inflorescences in line A was relatively higher thar in line B, probably due to the coincidence of the longest ascending period of radiation witl the peak of the flowering period in line A (Figure 5f).

In Figure 6, the number of ripened fruits and their weight are presented. There wer no statistically significant differences in the number and weight of fruits per plant betweer the two planting rows for each measurement during the experimental period. However, a the peak of the fruiting period, a trend for heavier fruits per plant was obvious (Figure 6a,b Shading reduced the number of fruits at the end of the experimental period by 20% and th weight of fruits by 45% (Figure 6c,d). The cumulative effect of shading was 10% and 35% for the number and weight of fruits, respectively (Figure 6e,f).

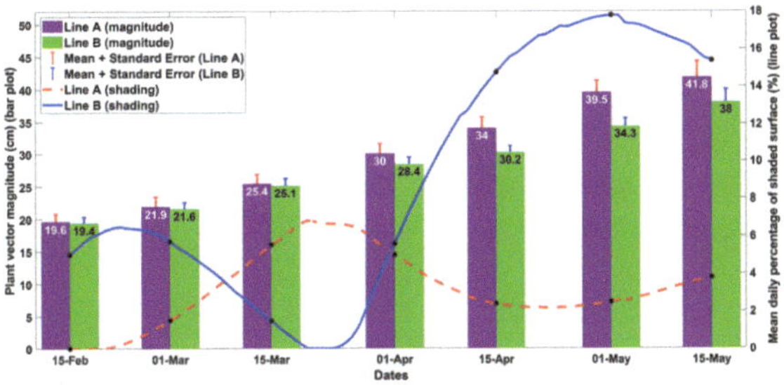

Figure 4. Mean plant vector magnitude for planting lines A and B (mean ± SE, n = 28) combined with the mean daily percentage of the shaded surface.

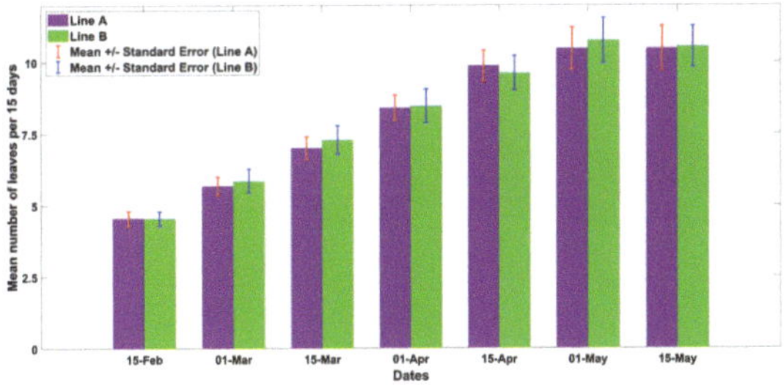

(a)

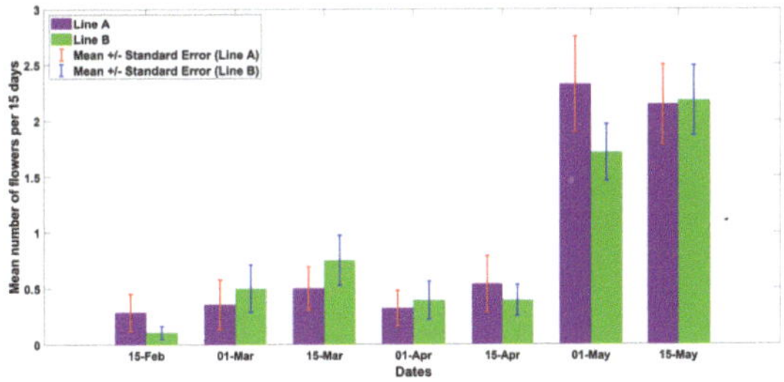

(b)

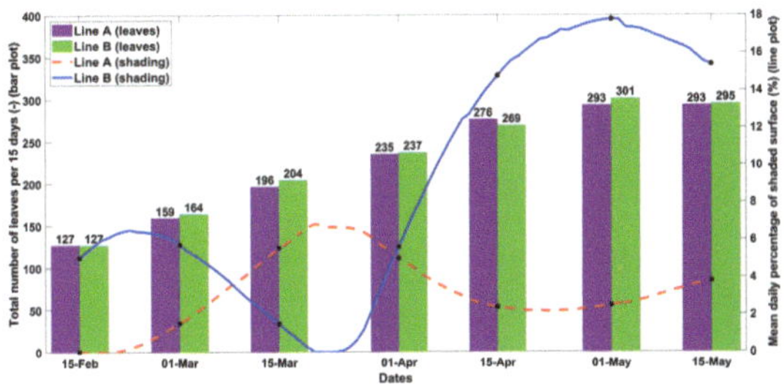

(c)

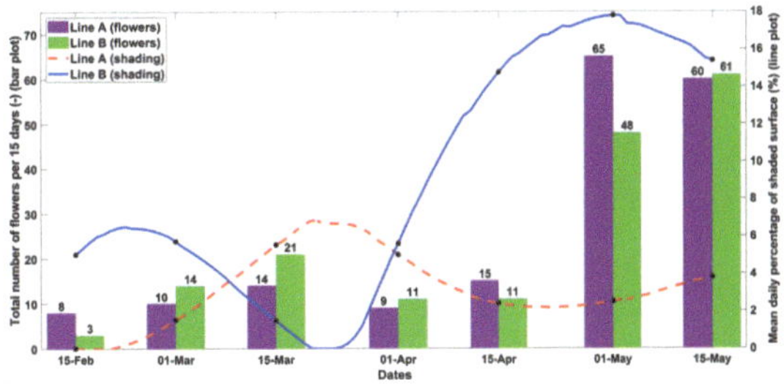

(d)

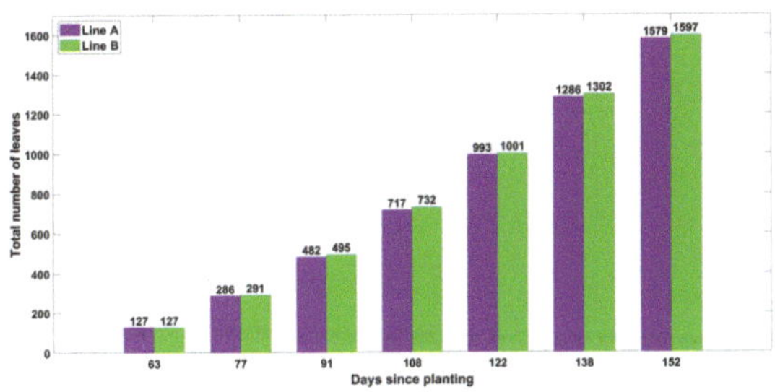

(e)

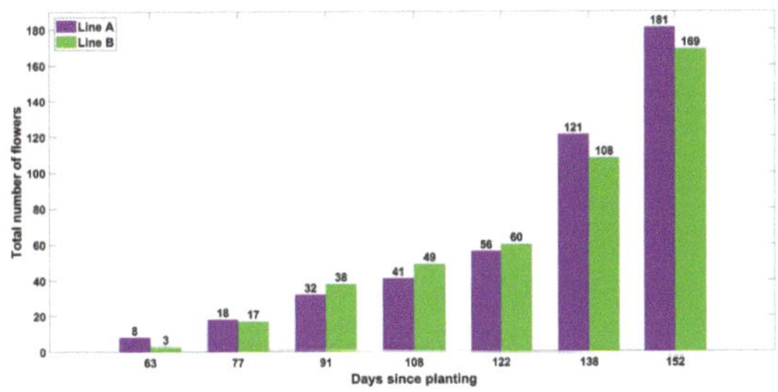

(f)

Figure 5. (**a**) Mean number of leaves during the experimental period (mean ± SE, n = 28). (**b**) Mean number of inflorescences during the experimental period (mean ± SE, n = 28). (**c**) Total number of leaves for each planting row combined with the mean daily percentage of the shaded surface. (**d**) Total number of inflorescences for each planting row combined with the mean daily percentage of the shaded surface. (**e**) Cumulative sum of the number of leaves for each planting row. (**f**) Cumulative sum of the number of inflorescences for each planting row.

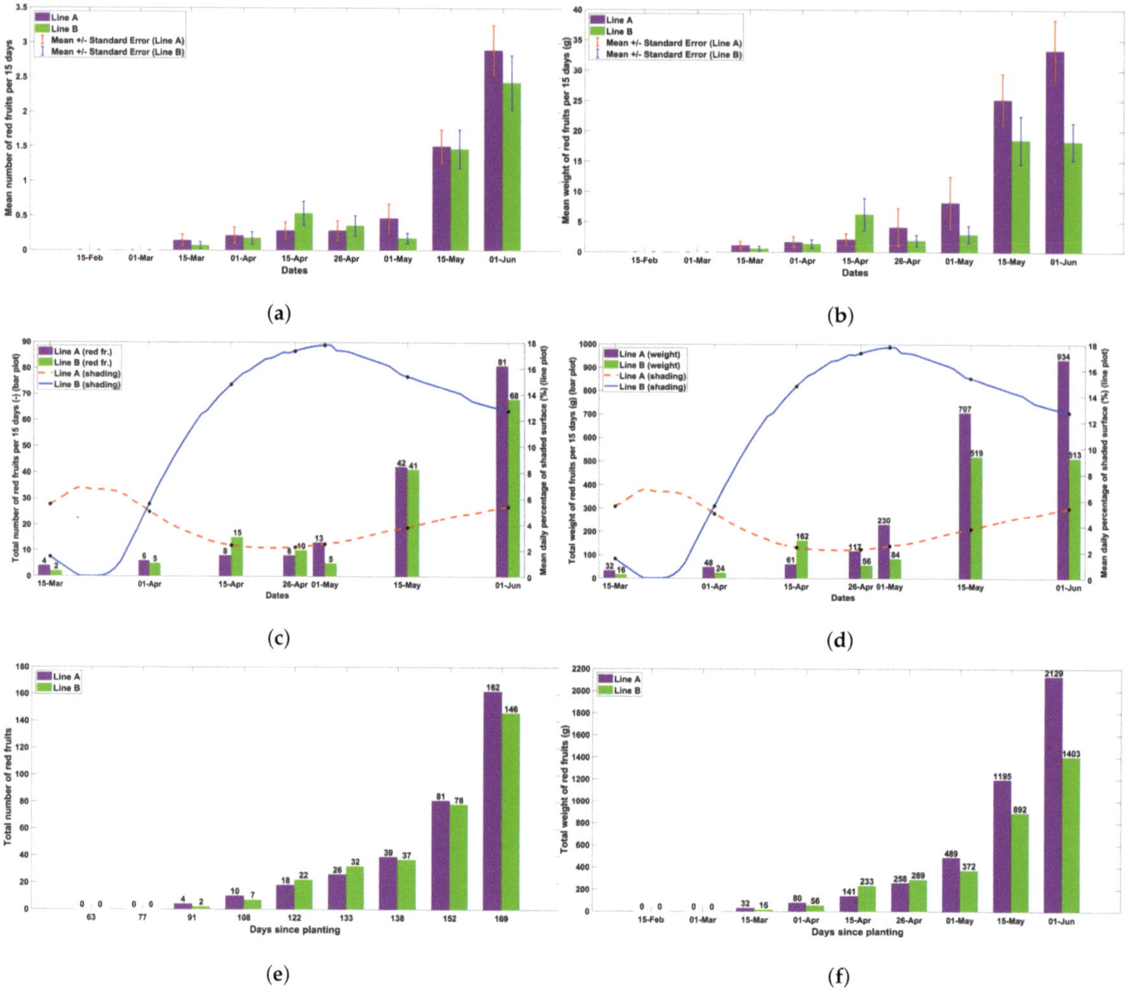

Figure 6. (**a**) Mean number of ripened (red) fruits during the experimental period (mean ± SE, n = 28). (**b**) Mean weight of ripened (red) fruits during the experimental period (mean ± SE, n = 28). (**c**) Total number of ripened fruits for each planting row combined with the mean daily percentage of the shaded surface. (**d**) Total weight of ripened fruits for each planting row combined with the mean daily percentage of the shaded surface. (**e**) Cumulative sum of the number of ripened fruits for each planting row. (**f**) Cumulative sum of the weight of ripened fruits for each planting row.

In Table 1, the number of leaves, flowers, ripened fruits, and weight of fruits at the final harvest are presented, as well as the percentage change between each parameter for lines A and B, together with the total shading posed by the photovoltaics on each planting line. The total shading values were calculated using the trapezoidal method with unit spacing, which computes the approximate integral of the data in Figure 2.

Table 1. Totals for the measured parameters and the percentage change between each parameter for lines A and B.

Parameter	Line A	Line B	Percentage Change (%)
Shading (m^2)	367.8	1099.6	+198.96
Number of Leaves	1579	1597	+1.14
Number of Flowers	181	169	−6.63
Number of Red Fruits	162	146	−9.88
Weight (g)	2123	1403	−33.91

3.2. Total Phenolic Content and Free-Radical-Scavenging Activity

As mentioned in the Materials and Methods Section, samples from each line underwent the TPC method for the determination of their total phenolic content. The absorbance of each sample was measured at 760 nm, three samples in triplicate, and their TPC is given through the calibration curve of GA. The results are given in Figure 7. The difference between the two lines of strawberries was significant, as line B had a significantly higher TPC content than line A.

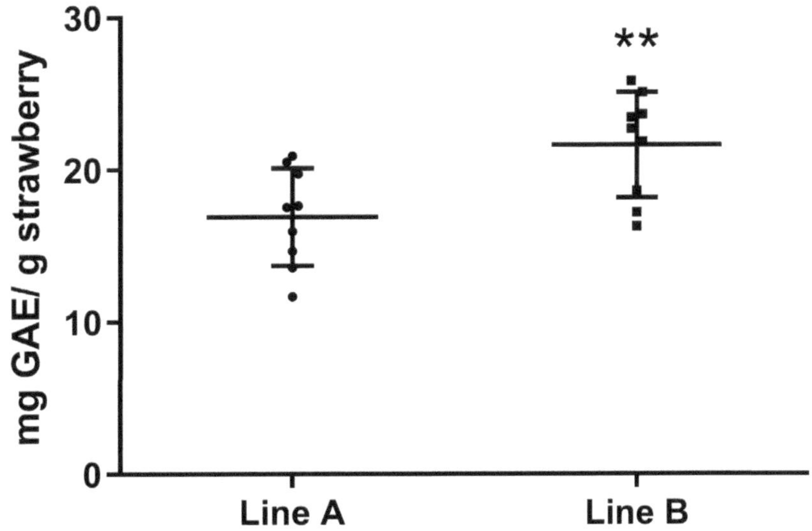

Figure 7. Results of total phenolic content in strawberries. The TPC in strawberries is expressed as mg of GAE per g of strawberries. Two asterisks (**) indicate statistically significant differences ($p < 0.01$) between the two strawberry planting lines.

Moreover, three samples from each line underwent the DPPH method for the determination of their free-radical-scavenging activity/antioxidant activity. The absorbance of each sample was measured at 517 nm in triplicate, and their scavenging potential is given through the calibration curve of L-ascorbic acid. The results are presented in Figure 8. The difference between the two groups of strawberries was significant, as line B had 1.375 ± 0.2524 mg of AAE per g morethan line A.

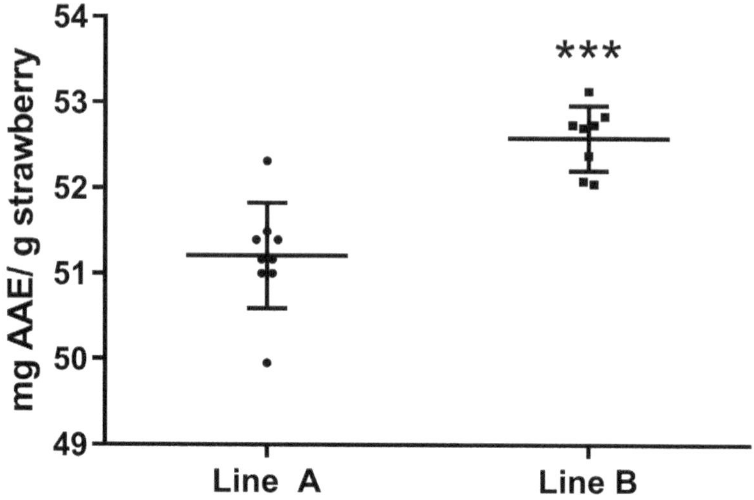

Figure 8. The results of the DPPH analysis in strawberries. The ability to scavenge DPPH i strawberries expressed as mg of AAE per g strawberry in different plot options. Three asterisks (** indicate statistically significant differences ($p < 0.0001$) between the two strawberry planting lines.

4. Discussion

Shading from photovoltaic arrays on the roof of greenhouses can have a positive o negative effect on the growth of the cultivated plants, depending on the period durin which the cultivation is carried out [11,33,34]. During the growth of strawberry plants i this study, due to diurnal and seasonal variations in light intensity and ambient temperatur plants may have experienced changes in incident radiation and, consequently, temperatur within the greenhouse daily, which included either quite low temperatures during th winter period or quite high temperatures during the spring period. Therefore, shadin from the PVs could theoretically have a negative effect on plant growth during the winte period, but mitigate the negative effect of high temperatures in the spring. At the sam time, reduced light intensities are expected to have a negative effect on plant growth [22 According to the results, there was a tendency of larger plants in the row with highe light intensities, while there was no effect of shading on the number of leaves throughou the experiment. This tendency of increased plant size was observed during spring, an therefore, this effect should be attributed to the slightly increased light intensity receive by plants in row A. At the same time, the prevailing air temperature for the two rows wa the same. Therefore, the effect of the photovoltaics on the shading was different betwee the two rows, but their effect on temperature was similar for all the plants. Consequentl any positive effect of the photovoltaics in mitigating high temperatures may contribut to the observed difference in plant size between the two rows, which otherwise could b smaller, if the more shaded plants had benefited from lower temperatures.

Shading did not affect the onset of flowering. The onset of flowering for each plan species is different and is usually determined by the prevailing photoperiod, i.e., the relativ ratio of light–dark hours in a day and the progressive decrease or increase of this ratio [35 This step in a plant's life cycle is significant due to its adaptability to seasonal changes an its reproductive success [36]. The onset of flowering is not affected by light intensity [35] which was confirmed in the present study, with the onset of flowering being the same i both crop lines regardless of the shading amount. In previous studies, the shading did no affect the number of flowers for pepper plants [37] and strawberry plants [38]. However, i our results, the number of flowers for each crop line was lower compared to the other, afte a preceding period of increasing shading. Since the total shading in line B was greater tha the shading in line A, the total number of flowers for line A was slightly higher than for line B at the end of the cultivation period.

The yield of fruits was affected by shading in terms of fruit weight. The number of fruits was not affected, except in the last harvest, when the line with the higher shading produced 20% fewer fruits. At the end of the experiment, the total number of fruits was slightly lower under the higher shading treatment. However, the difference in fruit weight was significantly higher between the two lines, because the strawberries under higher light intensity were about 50% heavier during the last month of the experiment, that is the period with the peak of fruiting. As the number of fruits was similar between the lines, it is obvious that, under higher light intensity, bigger fruits were produced. Smaller fruits under shading could result in an altered secondary metabolites profile and, thus, change the fruits' quality characteristics. In previous studies, the yield of tomatoes [39] was reduced only under the high level of shading (60%), while in a hydroponic culture of tomatoes, the reduction in the yield was gradually increased under increasing levels (15–50%) of shading [29]. On the other hand, the yield and quality of chili peppers improved under shading [37].

Shading is known to increase the TPC and antioxidant activity of many plants including coffee beans [40], perennial wall rocket [41], and sweet potato [42]. In our case, the results showed that the shading provided by the semi-transparent PV panels resulted in smaller fruits with a higher TPC and antioxidant activity.

Although the shading from the semi-transparent photovoltaics did not assist the production of large set fruits, even if the number was unaffected, the contribution of the shading resulted in higher quality fruits, while their wider impact in the energy part must not be neglected. Finally, it must be noted that the above results would probably be more intense if the comparison was between plants facing the increased shade from the photovoltaics and plants not receiving any additional shade at all.

In an era when efforts to create a complex between sustainable agriculture and RES are intensifying, the combination of greenhouse systems and semi-transparent photovoltaics can become a useful tool, not only for producing environmentally friendly energy, but also for qualitative optimization of greenhouse crops. The above example demonstrates how, through innovation, solutions may be discovered for two challenges: food security, on the one hand, and renewable energy production and climate change mitigation, on the other, for a more sustainable future.

Author Contributions: Conceptualization, T.P. and G.G.; methodology, T.P., P.I. and G.G.; software, T.P. and P.I.; validation, A.K., G.G. and N.K.; formal analysis, A.K., G.G. and N.K; investigation, T.P., P.I. and F.K.; resources, T.P., P.I. and F.K.; data curation, T.P., P.I. and F.K.; writing—original draft preparation, T.P., P.I. and G.G.; writing—review and editing, A.K. and N.K.; visualization, T.P. and P.I.; supervision, A.K. and N.K.; project administration, A.K.; funding acquisition, A.K. All authors have read and agreed to the published version of the manuscript.

Funding: The publication fees of this manuscript have been financed by the Research Council of the University of Patras.

Data Availability Statement: The data that support the findings of this study are available from the corresponding author upon reasonable request. The data are not publicly available due to confidentiality agreements with research participants.

Acknowledgments: The present work was financially supported by the "Andreas Mentzelopoulos Foundation". Special thanks to Sylvia Mangani for her invaluable contribution to the laboratory experiments conducted for this research.

Conflicts of Interest: The authors declare no conflicts of interest.

References

1. Khoshnevisan, B.; Rafiee, S.; Mousazadeh, H. Environmental impact assessment of open field and greenhouse strawberry production. *Eur. J. Agron.* **2013**, *50*, 29–37. [CrossRef]
2. Tang, Y.; Ma, X.; Li, M.; Wang, Y. The effect of temperature and light on strawberry production in a solar greenhouse. *Sol. Energy* **2020**, *195*, 318–328. [CrossRef]

3. Schneider, S.; Ziegler, C.; Melzer, A. Growth towards light as an adaptation to high light conditions in Chara branches. *Ne Phytol.* **2006**, *172*, 83–91. [CrossRef]
4. Tang, W.; Guo, H.; Baskin, C.C.; Xiong, W.; Yang, C.; Li, Z.; Song, H.; Wang, T.; Yin, J.; Wu, X.; et al. Effect of light intensity o morphology, photosynthesis and carbon metabolism of alfalfa (*Medicago sativa*) seedlings. *Plants* **2022**, *11*, 1688. [CrossRef]
5. Zivcak, M.; Brestic, M.; Kalaji, H.M.; Govindjee. Photosynthetic responses of sun- and shade-grown barley leaves to high light: the lower PSII connectivity in shade leaves associated with protection against excess of light? *Photosynth. Res.* **2014**, *119*, 339–35 [CrossRef]
6. Dong, C.; Fu, Y.; Liu, G.; Liu, H. Low light intensity effects on the growth, photosynthetic characteristics, antioxidant capacit yield and quality of wheat (*Triticum aestivum* L.) at different growth stages in BLSS. *Adv. Space Res.* **2014**, *53*, 1557–1566. [CrossRe
7. Zheng, Y.; Wang, S.Y.; Wang, C.Y.; Zheng, W. Changes in strawberry phenolics, anthocyanins, and antioxidant capacity i response to high oxygen treatments. *LWT Food Sci. Technol.* **2007**, *40*, 49–57. [CrossRef]
8. Miller, K.; Feucht, W.; Schmid, M. Bioactive compounds of strawberry and blueberry and their potential health effects based o human intervention studies: A brief overview. *Nutrients* **2019**, *11*, 1510. [CrossRef] [PubMed]
9. Skrovankova, S.; Sumczynski, D.; Mlcek, J.; Jurikova, T.; Sochor, J. Bioactive compounds and antioxidant activity in differe types of berries. *Int. J. Mol. Sci.* **2015**, *16*, 24673–24706. [CrossRef]
10. Choe, E.; Min, D.B. Mechanisms of antioxidants in the oxidation of foods. *Compr. Rev. Food Sci. Food Saf.* **2009**, *8*, 345–35 [CrossRef]
11. Touil, S.; Richa, A.; Fizir, M.; Bingwa, B. Shading effect of photovoltaic panels on horticulture crops production: A mini revie *Rev. Environ. Sci. Biotechnol.* **2021**, *20*, 281–296. [CrossRef]
12. Friman-Peretz, M.; Ozer, S.; Geoola, F.; Magadley, E.; Yehia, I.; Levi, A.; Brikman, R.; Gantz, S.; Levy, A.; Kacira, M.; et a Microclimate and crop performance in a tunnel greenhouse shaded by organic photovoltaic modules—Comparison wit conventional shaded and unshaded tunnels. *Biosyst. Eng.* **2020**, *197*, 12–31. [CrossRef]
13. Gorjian, S.; Calise, F.; Kant, K.; Ahamed, M.S.; Copertaro, B.; Najafi, G.; Zhang, X.; Aghaei, M.; Shamshiri, R.R. A review o opportunities for implementation of solar energy technologies in agricultural greenhouses. *J. Clean. Prod.* **2021**, *285*, 12480 [CrossRef]
14. Statuto, D.; Abdel-Ghany, A.M.; Starace, G.; Arrigoni, P.; Picuno, P. Comparison of the efficiency of plastic nets for shadin greenhouse in different climates. In *Innovative Biosystems Engineering for Sustainable Agriculture, Forestry and Food Productio* MID-TERM AIIA 2019; Coppola, A., Di Renzo, G., Altieri, G., D'Antonio, P., Eds.; Springer: Berlin/Heidelberg, Germany, 202 Volume 67, pp. 33–45. [CrossRef]
15. Castellano, S.; Mugnozza, G.S.; Russo, G.; Briassoulis, D.; Mistriotis, A.; Hemming, S.; Waaijenberg, D. Plastic nets in agricultur A general review of types and applications. *Appl. Eng. Agric.* **2008**, *24*, 799–808. [CrossRef]
16. Puglisi, R.; Lippolis, M.; Starace, G.; Arrigoni, P.; Picuno, P. Efficiency of plastic nets for greenhouse shading. In *AIIA 202 Biosystems Engineering towards the Green Deal*; Ferro, V., Giordano, G., Orlando, S., Vallone, M., Cascone, G., Porto, S.M.C., Eds Springer: Berlin/Heidelberg, Germany, 2023; Volume 337, pp. 125–134. [CrossRef]
17. Abdel-Ghany, A.; Al-Helal, I.; Alkoaik, F.; Alsadon, A.; Shady, M.; Ibrahim, A. Predicting the cooling potential of differer shading methods for greenhouses in arid regions. *Energies* **2019**, *12*, 4716. [CrossRef]
18. Santolini, E.; Pulvirenti, B.; Guidorzi, P.; Bovo, M.; Torreggiani, D.; Tassinari, P. Analysis of the effects of shading screens on th microclimate of greenhouses and glass facade buildings. *Build. Environ.* **2022**, *211*, 108691. [CrossRef]
19. Holcman, E.; Sentelhas, P. Microclimate under different shading screens in greenhouses cultivated with bromeliads. *Rev. Bra Eng. Agríc. Ambient.* **2012**, *16*, 858–863. [CrossRef]
20. Weselek, A.; Ehmann, A.; Zikeli, S.; Lewandowski, I.; Schindele, S.; Högy, P. Agrophotovoltaic systems: Applications, challenge and opportunities. A review. *Agron. Sustain. Dev.* **2019**, *39*, 35. [CrossRef]
21. Yano, A.; Cossu, M. Energy sustainable greenhouse crop cultivation using photovoltaic technologies. *Renew. Sustain. Energy Re* **2019**, *109*, 116–137. [CrossRef]
22. Ahmed, H.A.; Al-Faraj, A.A.; Abdel-Ghany, A.M. Shading greenhouses to improve the microclimate, energy and water saving i hot regions: A review. *Sci. Hortic.* **2016**, *201*, 36–45. [CrossRef]
23. Hassanien, R.H.E.; Li, M.; Yin, F. The integration of semi-transparent photovoltaics on greenhouse roof for energy and plar production. *Renew. Energy* **2018**, *121*, 377–388. [CrossRef]
24. Angmo, P.; Phuntsog, N.; Namgail, D.; Chaurasia, O.P.; Stobdan, T. Effect of shading and high temperature amplitude i greenhouse on growth, photosynthesis, yield and phenolic contents of tomato (*Lycopersicum esculentum* Mill.). *Physiol. Mol. Bic Plants* **2021**, *27*, 1539–1546. [CrossRef]
25. Milenković, L.; Mastilović, J.; Kevrešan, V.; Bajić, A.; Gledić, A.; Stanojević, L.; Cvetković, D.; Šunić, L.; Ilić, Z.S. Effect of shadin and grafting on yield and quality of tomato. *J. Sci. Food Agric.* **2020**, *100*, 623–633. [CrossRef]
26. Wang, J.; Wang, H. Effects of shade on strawberries in hydroponic cultivation. *Acta Hortic.* **2014**, *1049*, 733–736. [CrossRef]
27. Blando, F.; Gerardi, C.; Renna, M.; Castellano, S.; Serio, F. Characterisation of bioactive compounds in berries from plants grow under innovative photovoltaic greenhouses. *J. Berry Res.* **2018**, *8*, 55–69. [CrossRef]
28. Kavga, A.; Trypanagnostopoulos, G.; Zervoudakis, G.; Tripanagnostopoulos, Y. Growth and physiological characteristics c lettuce (*Lactuca sativa* L.) and rocket (*Eruca sativa* Mill.) plants cultivated under photovoltaic panels. *Not. Bot. Horti Agrobot.* **201** *46*, 206–212. [CrossRef]

9. López-Díaz, G.; Carreño-Ortega, A.; Fatnassi, H.; Poncet, C.; Díaz-Pérez, M. The effect of different levels of shading in a photovoltaic greenhouse with a North–South orientation. *Appl. Sci.* **2020**, *10*, 882. [CrossRef]
0. Petrakis, T.; Thomopoulos, V.; Kavga, A.; Argiriou, A.A. An algorithm for calculating the shade created by greenhouse integrated photovoltaics. *Energy Ecol. Environ.* **2023**. [CrossRef]
1. Ainsworth, E.; Gillespie, K. Estimation of total phenolic content and other oxidation substrates in plant tissues using Folin–Ciocalteu reagent. *Nat. Protoc.* **2007**, *2*, 875–877. [CrossRef] [PubMed]
2. Golpour, I.; Ferrão, A.C.; Gonçalves, F.; Correia, P.M.; Blanco-Marigorta, A.M.; Guiné, R.P. Extraction of phenolic compounds with antioxidant activity from strawberries: Modelling with artificial neural networks (ANNs). *Foods* **2021**, *10*, 2228. [CrossRef] [PubMed]
3. Kavga, A.; Trypanagnostopoulos, G.; Koulopoulos, A.; Tripanagnostopoulos, Y. Implementation of photovoltaics on greenhouses and shading effect to plant growth. *Acta Hortic.* **2019**, *1242*, 749–756. [CrossRef]
4. Sarıdaş, M.A. Seasonal variation of strawberry fruit quality in widely grown cultivars under Mediterranean climate condition. *J. Food Compos. Anal.* **2021**, *97*, 103733. [CrossRef]
5. Erwin, J. Factors affecting flowering in ornamental plants. In *Flower Breeding and Genetics*; Anderson, N.O., Ed.; Springer: Dordrecht, The Netherlands, 2007; pp. 1–13. [CrossRef]
6. Bäurle, I.; Dean, C. The timing of developmental transitions in plants. *Cell* **2006**, *125*, 655–664. [CrossRef]
7. Hassanien, R.H.E.; Ibrahim, M.M.; Ghaly, A.E.; Abdelrahman, E.N. Effect of photovoltaics shading on the growth of chili pepper in controlled greenhouses. *Heliyon* **2022**, *8*, e08877. [CrossRef]
8. Hassanien, R.H.E.; Ming, L. Influence of opaque photovoltaic shading on microclimate and growth of strawberry in greenhouses. *Misr J. Agric. Eng.* **2021**, *38*, 377–390. [CrossRef]
9. Callejón-Ferre, Á.J.; Manzano-Agugliaro, F.; Díaz-Pérez, M.; Carreño-Ortega, A.; Pérez-Alonso, J. Effect of shading with aluminised screens on fruit production and quality in tomato (*Solanum lycopersicum* L.) under greenhouse conditions [Efecto del sombreo mediante pantallas aluminizadas sobre la producción y calidad de fruto en tomate (*Solanum lycopersicum* L.) bajo invernadero]. *Span. J. Agric. Res.* **2009**, *7*, 41–49. [CrossRef]
0. Somporn, C.; Kamtuo, A.; Theerakulpisut, P.; Siriamornpun, S. Effect of shading on yield, sugar content, phenolic acids and antioxidant property of coffee beans (*Coffea arabica* L. cv. Catimor) harvested from north-eastern Thailand. *J. Sci. Food Agric.* **2012**, *92*, 1956–1963. [CrossRef] [PubMed]
1. Caruso, G.; Formisano, L.; Cozzolino, E.; Pannico, A.; El-Nakhel, C.; Rouphael, Y.; Tallarita, A.; Cenvinzo, V.; De Pascale, S. Shading Affects Yield, Elemental Composition and Antioxidants of Perennial Wall Rocket Crops Grown from Spring to Summer in Southern Italy. *Plants* **2020**, *9*, 933. [CrossRef] [PubMed]
2. Jing, X.; Chen, P.; Jin, X.; Lei, J.; Wang, L.; Chai, S.; Yang, X. Physiological, Photosynthetic, and Transcriptomics Insights into the Influence of Shading on Leafy Sweet Potato. *Genes* **2023**, *14*, 2112. [CrossRef] [PubMed]

plants

MDPI

Article

Optimized Design of Irrigation Water-Heating System and Its Effect on Lettuce Cultivation in a Chinese Solar Greenhouse

Liangjie Guo [1], Xinyi Chen [1], Shiye Yang [1], Ruimin Zhou [1], Shenyan Liu [1] and Yanfei Cao [1,2,*]

[1] College of Horticulture, Northwest A & F University, Yangling 712100, China; glj18503443835@163.com (L.G.); chenxyiii@nwafu.edu.cn (X.C.); ysy0819@nwafu.edu.cn (S.Y.); zhouruimin@nwafu.edu.cn (R.Z.); liuas@nwafu.edu.cn (S.L.)

[2] Key Laboratory of Protected Horticultural Engineering in Northwest, Ministry of Agriculture and Rural Affairs, Northwest A&F University, Yangling 712100, China

* Correspondence: caoyanfei@nwsuaf.edu.cn

Abstract: In cold regions, the low irrigation water temperature is an important factor of lov temperature stress for greenhouse crops. In this paper, an irrigation water-heating system (IWHS) proposed to increase the water temperature by utilizing the excess heat in the solar greenhouse. Th heat-collection capacity of the system was analyzed by screening the IWHS process parameters in Chinese solar greenhouse, and a warm-water irrigation experiment for lettuce was conducted. Th results demonstrated that the water temperature increased with the increase in wind speed, and th increase in daily average water temperature reached the maximum value of 8.6 °C at 4.5 m/s win speed. When the heat exchanger was installed at a height of 3.0 m, the collector capacity increase by 17.8% and 6.0% compared with the heating capacity at 0 m and 1.5 m, respectively, and th operation termination water temperature was 22.0–32.2 °C and its coefficient of performance (COI was optimal. Surface darkening of the heat exchanger did not affect the heat-collection capacity c the system. Using the IWHS effectively improved the temperature of lettuce irrigation water in th Chinese solar greenhouse. The increased frequency of warm-water irrigation significantly promote lettuce growth and increased the average yield per plant by 15.9%. Therefore, IWHS effectivel increased the irrigation water temperature in a Chinese solar greenhouse in winter. Improving th system would enhance its economic and application value.

Keywords: irrigation water-heating system; Chinese solar greenhouse; warm-water irrigation; lettuc

Citation: Guo, L.; Chen, X.; Yang, S.; Zhou, R.; Liu, S.; Cao, Y. Optimized Design of Irrigation Water-Heating System and Its Effect on Lettuce Cultivation in a Chinese Solar Greenhouse. *Plants* **2024**, *13*, 718. https://doi.org/10.3390/plants13050718

Academic Editors: Weijie Jiang and Wenna Zhang

Received: 4 February 2024
Revised: 26 February 2024
Accepted: 27 February 2024
Published: 4 March 2024

1. Introduction

Protected horticulture has developed rapidly in China, the Netherlands, Israel, th United States, Spain, and other countries over the past 60 years, becoming an impo tant national economic industry [1]. By 2022, the total horticultural area in China wa >2.8 million ha, accounting for >80% of the total protected horticulture area globally. Ch nese solar greenhouses accounted for 29% of the total Chinese horticultural area, effectivel ensuring an annual stable supply of agricultural products, such as vegetables, fruits, anc flowers, in northern China [2]. Areas of northern China that feature a cold climate result i winter irrigation water being colder than the minimum irrigation temperature required fo plant growth [3]. Furthermore, practical irrigation water-warming measures in actual pro duction are lacking. Therefore, improving irrigation water-heating technology to solve th issue of low-temperature stress of Chinese solar greenhouse crops and improve greenhous production efficiency is of great significance to horticultural production in cold areas.

Temperature is the microclimate factor that affects crop growth the most in Chines solar greenhouses [4], and root zone temperature has a greater effect on plant stress thar air temperature [5]. Numerous studies have been conducted on root zone temperature reg ulation methods and strategies [6–11]. Furthermore, some studies have analyzed the effect of root zone temperature on crop physiology and ecology [12–15], nutrient absorption [16] and yield [15]. Some studies confirmed that root zone temperature regulation positively affected crop growth [17–19], and the irrigation water temperature was one of the main factors affecting root zone temperature. Several researchers mainly studied the influence of warm-water irrigation on flowers and field crops [20,21], and confirmed the significance of warm-water irrigation for crop growth. For vegetable facilities, warm-water irrigation research was mainly conducted on leaf vegetable crops [22]. Hooks et al. conducted a hydroponic experiment on lettuce in winter [23], and reported that a nutrient solution heating temperature of 22 °C increased the yield by 31.4% compared with no heating, indicating that increasing the winter root zone temperature positively affected lettuce growth. The appropriate water supply is the crucial factor for obtaining a high yield and quality of vegetable crops [24,25], as water regulates the physiological and biochemical state of the plant under both normal and stressful conditions [26].

The irrigation water-heating measures widely used in production mainly include solar water pools [27] and burning non-renewable resources [28]. Although the above heating methods have many applications, they are subject to limitations. For example, building a solar water pool in a solar greenhouse requires a certain amount of greenhouse land. Additionally, the tight coverage of the pool raises the water temperature slowly, and the increase is small, which cannot meet irrigation requirements. The coverage is not exact, and increases the greenhouse air humidity, which easily induces high-humidity diseases. Some heating devices heat irrigation water through electric heating or burning non-renewable resources. Although the heating capacity is strong and the heating speed is fast, the cost is high, which is not conducive to energy saving and environmental protection. Furthermore, the irrigation water-heating system (IWHS) based on solar photovoltaic/thermal technology [29] has good overall performance but high investment and maintenance costs.

The greenhouse air temperature is relatively high at approximately noon in winter, and contains abundant air heat energy [30]. Many researchers have used water as a medium to collect air excess heat and solar energy and have designed various greenhouse heat-collection devices [31,32]. The air excess heat heating technology of solar greenhouses includes the Earth–gas heat exchange system [33] and air waste heat pump technology [34], which can store excess greenhouse heat during the day through "peak cutting and valley filling" and is used to increase the root zone temperature and air temperature. Based on the above research, this experiment used greenhouse air excess heat and insulation technology to improve the solar greenhouse irrigation water temperature in winter and spring to develop a suitable IWHS.

Therefore, this study proposed an IWHS. This experiment screened the wind speed, height, and other process parameters of the heat exchange device, and analyzed the system application effect on lettuce to determine the best IWHS process parameters and explore the effects of the IWHS on lettuce growth. It is hoped that a new method and practical reference for warm-water irrigation of solar greenhouse lettuce in winter will be established.

2. Results and Analysis

2.1. *Effects of Different Wind Speeds on the Heating Effect of the System*

Wind speed is an important factor affecting heat transfer, and the appropriate wind speed determines the heat-collection performance of the system. This experiment used three wind speeds in the screening test. The heat-collection capacity of the system increased with the increase in wind speed when the temperature was higher on five sunny days, and the run time was 3.0 h. At wind speeds of 2.3, 3.4, and 4.5 m/s, the average daily increase was 7.7, 8.3, and 8.6 °C, respectively (Table 1). Compared with the wind speeds of 2.3 and 3.4 m/s, the 4.5 m/s wind speed increased the water temperature by 11.7% and 3.6%, respectively. The faster wind speed increased the system ventilation volume, promoted the rapid exchange of more heat, and improved the system heating capacity.

Table 1. The effects of different wind speeds on the heating effect of IWHS.

Date	Working Time	Initial Water Temperature (°C)~ Final Water Temperature (°C)		
		2.3 m/s	3.4 m/s	4.5 m/s
5 December	11:30~14:30	9.4~17.0	9.5~17.5	9.3~17.5
8 December	10:30~13:30	10.6~17.5	10.5~17.8	10.4~18.3
12 December	11:00~14:00	11.5~20.9	11.6~21.6	11.7~22.1
13 December	10:50~13:50	12.4~18.7	12.3~19.1	12.2~19.3
14 December	10:20~13:20	10.9~19.2	10.8~20.0	10.8~20.4

2.2. Effects of Heat Exchangers of Different Heights on the Heating Effect of the System

Due to the influence of the solar greenhouse back roof and insulation quilt, the distribution of solar radiation and air temperature at different heights near the back wall of the greenhouse varied. Therefore, it was important to screen the heat exchanger height. When the system ran for 4.5~6.0 h in the maximum daily temperature period, the heat-collection performance of the system was $T_{h3} > T_{h2} > T_{h1}$. The T_{h1}, T_{h2}, and T_{h3} water temperature increased by an average of 9.0, 10.0, and 10.6 °C, respectively (Table 2). The T_{h3} water temperature increased by 17.8% and 6.0%, respectively, compared with that of Th1 and T_{h2}, and the T_{h3} water temperature at the end of the operation could reach 22.0~32.2 °C. On 31 October 2022, reducing greenhouse ventilation achieved a maximum water temperature of 32.2 °C. Thus, the system heat-collection efficiency was improved by reducing ventilation without affecting normal crop growth. However, the T_{h1} total solar radiation was 2.4 and 1.2 times that of T_{h3} and T_{h2}, respectively, and the system heat-collection ratio was opposite to that of the heat system collection, indicating that solar radiation did not significantly affect the system heat collection. The air inlet temperature at different heights was monitored on 19 November. The average inlet air temperatures of T_{h1}, T_{h2}, and T_{h3} were 22.2, 22.9, and 23.4 °C, respectively, which positively correlated with the temperature rise. Therefore, the heat exchanger placement height should be determined according to the air temperature distribution and plant height and should be placed in the middle and upper part of the ridge height direction.

Table 2. Effect of different heights of heat exchangers on the heating effect of IWHS.

Data	Working Time	Initial Water Temperature (°C)~ Final Water Temperature (°C)			Total Amount of Solar Radiation (MJ)			η		
		Bottom T_{h1}	Middle T_{h2}	Top T_{h3}	Bottom T_{h1}	Middle T_{h2}	Top T_{h3}	Bottom T_{h1}	Middle T_{h2}	Top T_{h3}
31 October	8:30~14:30	16.6~29.9	16.6~31.5	16.5~32.2	6.9	5.6	2.3	0.8	1.1	2.7
12 November	10:00~14:30	14.7~21.7	14.3~21.7	14.3~22.0	3.9	3.3	1.5	0.7	0.9	2.0
14 November	9:30~14:30	13.9~22.7	13.7~23.5	13.6~24.2	7.4	6.2	3	0.5	0.6	1.4
15 November	9:00~14:30	14.7~21.6	14.1~22.2	14.1~22.6	6.4	5.4	3.1	0.4	0.6	1.1
19 November	9:00~14:00	13.5~22.3	13.0~22.8	13.0~23.5	7.4	5.9	3.4	0.5	0.7	1.2

Note: T_{h1}, the heat exchanger is 0 m from the ground; T_{h2}, the heat exchanger is 1.5 m from the ground; T_{h3}, the heat exchanger is 3.0 m from the ground.

2.3. Effects of Heat Exchanger Surface Darkening on the System Heating Effect

Figure 1 depicts the system water temperature change after the heat exchanger surface was darkened. There was a small difference (0.1 °C) in the water temperatures between the darkened and undarkened surfaces. This analysis revealed that the low air temperature at the fan outlet resulted in a low device surface temperature, and the water did not effectively absorb the solar radiation.

2.4. Changes in IWHS Water Temperature in Lettuce Greenhouses

The water temperature changes of the IWHS in the lettuce greenhouse during four consecutive sunny days (28 January to 31 January 2023) were selected for analysis. Figure demonstrates that the minimum and maximum greenhouse indoor air temperatures during this period were 3.8 °C and 29.0 °C, respectively, and there was a period when the indoor temperature was lower than the minimum growth temperature requirement of lettuce. In the morning, the indoor air temperature rises. The IWHS water temperature gradually rises when the IWHS meets the control strategy and begins operating. The water temperature increase rate slows when the difference between the air and water temperatures decreases. The average IWHS maximum water temperature for four days was 23.2 °C, reaching the suitable water temperature for lettuce irrigation, which is 11.2 °C higher than that of unheated water. Due to normal greenhouse ventilation, the indoor temperature was above 26.0 °C for a short time at noon, and IWHS heating capacity was limited. The indoor temperature decreased as the solar radiation intensity decreased, and the system stopped operating. When the IWHS stopped, the water temperature decreased by 2.8~3.1 °C per day, indicating that the thermal insulation tank capacity was limited. The thermal insulation capacity can be improved by increasing the thermal insulation cotton thickness.

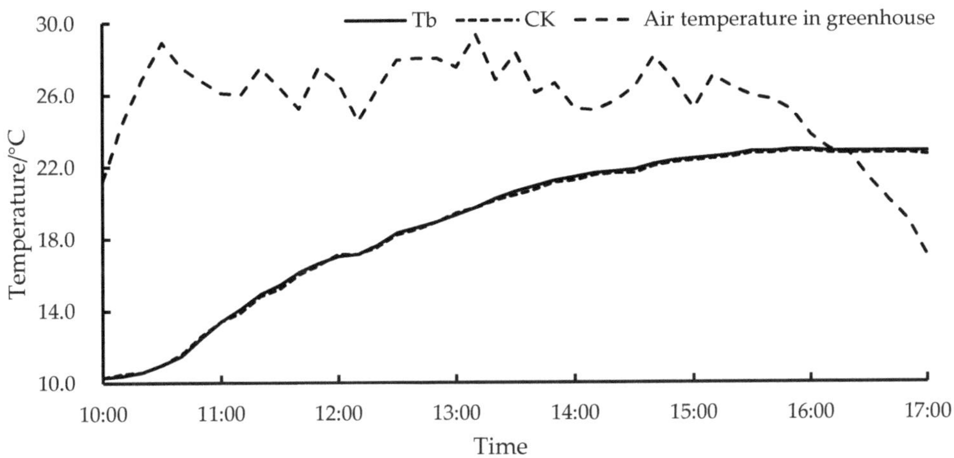

Figure 1. The system water temperature change after the heat exchanger surface was darkened. Tb, the system with darkened heat exchanger; CK, the system with undarkened heat exchanger.

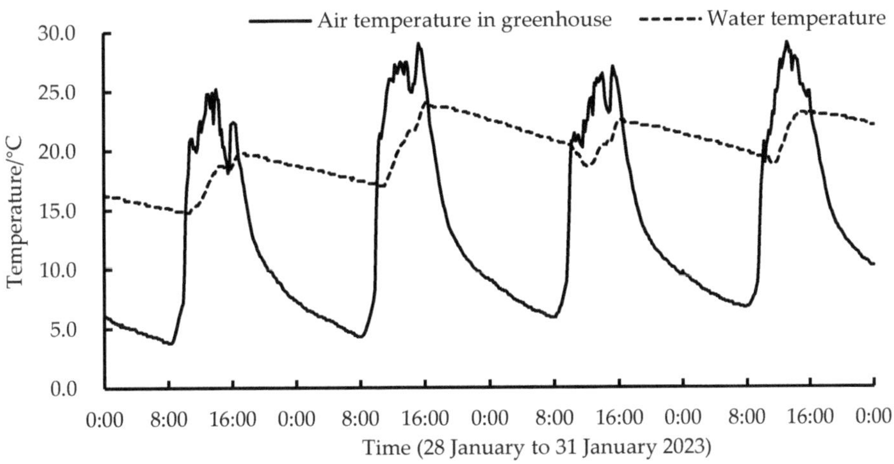

Figure 2. Changes in IWHS water temperature in lettuce greenhouses under sunny conditions.

2.5. Heat-Collection Performance of IWHS in Lettuce Greenhouse

The IWHS heat-collection capacity and overall energy consumption were calculate using the test data of four sunny days during the test period (Table 3). The system hea collection capacity was 5.60~10.90 MJ and the average COP was 3.07. On 19 January 202 the initial water temperature was higher and the system heat collection was lower due t the higher air temperature that day resulting in a higher COP than that in other periods.

Table 3. IWHS heat-collection performance of lettuce greenhouse.

Date	Working Time	Working Hours (h)	Initial Water Temperature (°C)~Final Water Temperature (°C)	Water Temperature Rise (°C)	Heat Collection (MJ)	COP
8 December	11:00~14:30	3.50	14.4~23.1	8.7	10.90	2.06
29 December	12:42~15:12	2.50	11.8~20.3	8.5	10.60	2.96
7 January	14:10~16:20	2.17	11.9~19.2	7.3	9.10	2.79
19 January	12:50~13:40	0.83	18.0~22.5	4.5	5.60	4.47

2.6. Effects of Warm-Water Irrigation on Lettuce Soil Temperature

Figure 3 depicts the effects of warm-water irrigation on the lettuce soil temperatur The soil temperature data during the irrigation period from 19 January to 20 January 202 were analyzed. Figure 3 demonstrates that warm-water irrigation increased the lettuc soil temperature by 4.0 °C at most, and the average soil temperature of the experimenta group was 1.2 °C higher than that of the control group within 6.0 h of irrigation treatmen Subsequently, the influence of irrigation water temperature on soil temperature gradu ally decreased as the irrigation treatment time increased, and the temperature differenc between the experimental and control groups gradually became consistent.

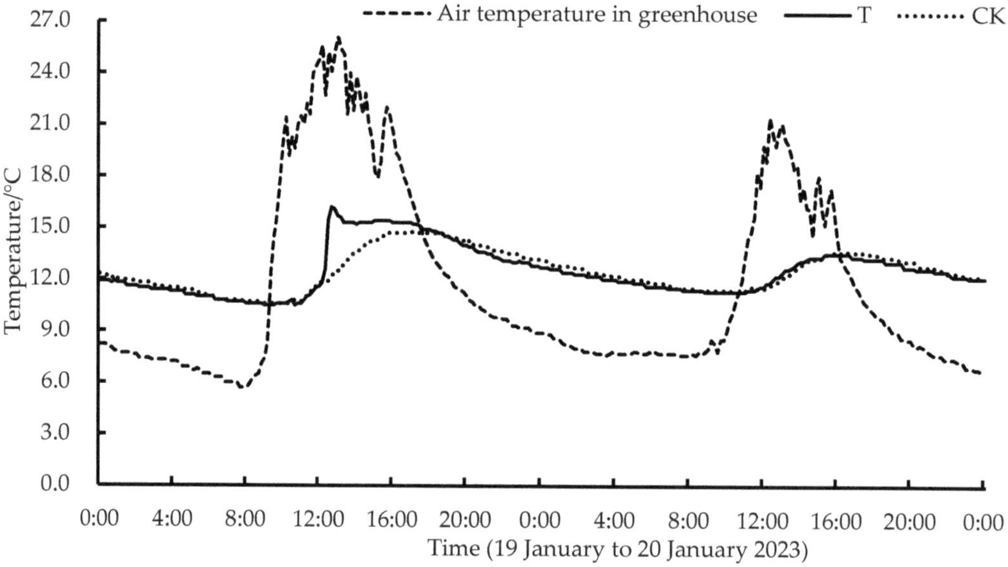

Figure 3. Effects of warm-water irrigation on lettuce soil temperature. T, the soil temperature i warm-water irrigation area; CK, the soil temperature in normal water irrigation area.

2.7. Effects of Warm-Water Irrigation on Lettuce Growth Indexes

Figure 4 depicts the changes in lettuce plant height and stem diameter after warr water irrigation. The control area (CK) and test area (T) plant heights exhibited significan differences after 35 days of warm-water irrigation. At harvest, the CK and T plant heights were 20.33 cm and 23.26 cm, respectively, increasing by ~9.6%. The CK and T stem diameters exhibited significant differences after 42 days of warm-water irrigation. At harvest, the CK and T stem diameters were 17.07 mm and 17.89 mm, respectively, an increase of ~4.8%. The results indicated that warm-water irrigation significantly promoted the growth of lettuce plant height and stem diameter and weakened the stress effect of low temperature on lettuce growth.

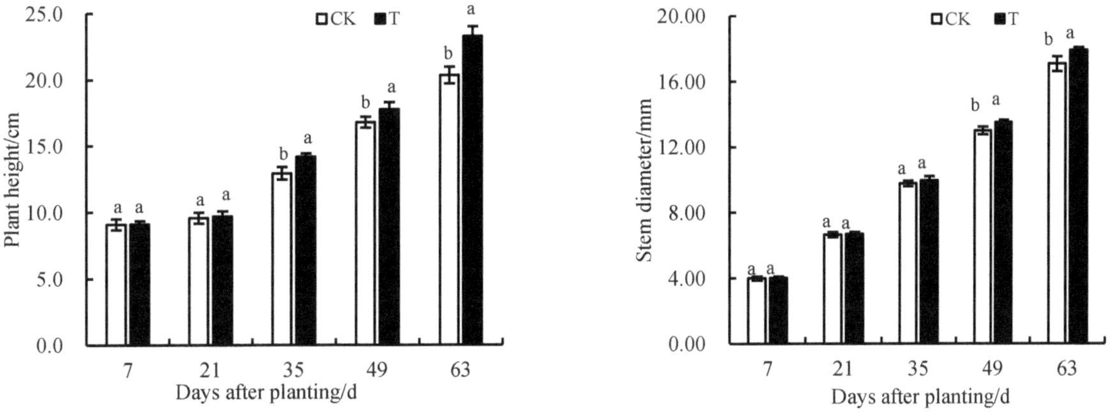

(a) (b)

Figure 4. (**a**) Effects of warm-water irrigation on lettuce plant height of Lettuce. (**b**) Effects of warm-water irrigation on lettuce stem diameter. T, the experimental group plants; CK, the control group plants. Lowercase letters (a and b) indicate statistical significance by Duncan's multiple range test ($p < 0.05$).

2.8. Effects of Warm-Water Irrigation on Lettuce Root and Leaf Indexes

Table 4 reports the effects of warm-water irrigation on the lettuce root index. The warm-water irrigation significantly increased the lettuce root surface area, the number of root tips, the number of leaves, and the leaf area, and T increased by 16.4%, 14.3%, 8.7%, and 16.7%, respectively, compared with CK. The results demonstrated that warm-water treatment significantly promoted lettuce root and leaf growth, and improved nutrient and water absorption.

Table 4. Effects of warm-water irrigation on lettuce root and leaf indexes.

Treatment	Root Surface Area (cm^2)	Root Tips (No.)	Leaves (No.)	Leaf Area (cm^2)
CK	17.07 ± 0.90 b	757.80 ± 26.83 b	19.71 ± 0.76 b	296.34 ± 17.00 b
T	19.87 ± 1.08 a	866.00 ± 59.86 a	21.43 ± 1.13 a	345.78 ± 14.78 a

Note: T, the experimental group plants; CK, the control group plants. Lowercase letters (a and b) indicate statistical significance by Duncan's multiple range test ($p < 0.05$).

2.9. Effects of Warm-WaterIrrigation on Lettuce Yield

Table 5 reports the effects of warm-water irrigation on lettuce yield. Warm-water irrigation significantly increased the yield per lettuce plant, which was increased by 15.9% compared with the control group. The results demonstrated that warm-water treatment significantly increased the lettuce yield and farmers' income.

Table 5. Effects of warm-water irrigation on the yield of lettuce.

Treatment	Yield per Plant (g)
CK	238.76 ± 16.98 b
T	276.62 ± 9.99 a

Note: T, the experimental group plants; CK, the control group plants. Lowercase letters (a and b) indicate statistic significance by Duncan's multiple range test ($p < 0.05$).

3. Discussion

In this experiment, continuous monitoring of the water temperature of a small sola water bucket revealed that it had poor heating capacity and required a long duratio while nighttime and cloudy days decreased the water temperature rapidly to the initia water temperature. Yue et al. [35] used a large solar water bucket, and only raised th water temperature for irrigation in the entire production period of strawberries by 1~2 °C which also indicated that the solar water bucket had a poor heating capacity. The IWH rapidly increased the water temperature for irrigation, and the water storage barrel has thermal insulation effect in low-temperature environments with less heat loss. The proces parameter screening experiment of the system determined that the faster wind spee increased the system ventilation volume, promoted the rapid exchange of more heat, an improved the system heating capacity. However, higher wind speeds result in redundar work by the system, reducing the system coefficient of performance (COP) and adversel affecting normal crop growth [36]. Therefore, the experiment did not conduct a larger win speed optimization test. The total solar radiation was not positively correlated with th system heat-collection efficiency, but was directly proportional to the air temperature of th fan air inlet. The analysis revealed that the heat exchanger surface temperature was lo due to the cold wind blowing out after heat exchange with cold water on the outer surfac of the heat exchanger, and the heat absorbed by solar radiation on the heat exchange surface was rapidly lost without exchanging with water. Even darkening the device surfac did not increase the system surface temperature to allow heat exchange with water.

Preliminary testing of the IWHS heat-collection performance revealed that the lettuc greenhouse system COP was 3.07, while Song et al. [37] reported that the average COP c the thermal collecting and releasing system developed with fan-coil units was 7.5. The mai reason for this was that different amounts of water were heated, and more water coul maintain a larger difference between water temperature and air temperature, resulting i higher heat-collection efficiency. In the initial heat-collection stage, the test system COP wa also >10. The heat-collection efficiency decreased as the temperature difference betwee water and air decreased, and the COP gradually decreased, resulting in a low overa heat-collection COP. Furthermore, the maximum greenhouse air temperature during hea collection was 27.4 °C, which was lower than the 33.7 °C of the greenhouse reported b Song et al. [37]. Additionally, the water flow rate positively correlated with the syster heat-collection performance. A larger water flow rate resulted in a larger heat-collectio power and COP of the system. The maximum water flow rate in this test was 0.59 m/ which was lower than that of the other side (1.2 m/s). The analysis of the application effe of warm-water irrigation on lettuce determined that the test results were consistent wit previous studies that used warm-water irrigation to promote tomato and cabbage growt These analysis findings may be because warm-water irrigation promotes the absorptio of soil nutrients by plants, thereby promoting growth [16]. Furthermore, the influence c warm-water irrigation on the soil temperature of lettuce was consistent with the study c Deng et al. [38].

Previous studies reported that the optimal COP of electric hot water was ~0.9. 100 m^3 of irrigation water must be heated by 15 °C during the winter production perio the electric heating method requires 1933 KW·h electricity, while the IWHS only require 567 KW·h electricity. The agricultural electricity price in Shaanxi Province is approximatel 0.5 RMB/(KW·h), implying savings of 683 RMB. The system heat-collection efficiency can b improved by reducing the greenhouse ventilation duration to reduce heat loss and improv the device performance without affecting normal crop growth. Due to the relatively low irrigation frequency in winter, subdivision irrigation can also reduce equipment input. The system fan has an active air exchange function, and the heat exchanger has a heat release function. Subsequent research can integrate the heat collection, heat release, and active dehumidification functions of the fan to improve its functionality and economy.

Although a suitable irrigation water temperature benefits plant growth, a new device application necessitates systematic research, establishing relevant basic databases for different plants, growth stages, and environmental temperatures, and studying the most energy-efficient operation strategy and most efficient irrigation measures to improve the system application effect. Follow-up research should examine the temperature field and water spatial distribution characteristics of the root zone under warm-water irrigation for different cultivation modes, such as soil, substrate, and coconut bran cultivation, and establish the corresponding mathematical models. In addition, we also need to study optimal control strategies for root zone temperature and air temperature to provide a more suitable growth environment for plants, reduce the impact of low temperature stress on plant growth, and promote increased yields and income [39].

4. Materials and Methods

4.1. The IWHS

The IWHS comprises a heat exchanger, a control electric box, water circulation pipes, a water pump, and a heat preservation storage bucket (Figure 5). The water in the heat exchanger flows from the bottom mouth to the upper mouth. In the high temperature period in the solar greenhouse, the fan and circulating water pump operate simultaneously under the condition of meeting the control strategy. As the air temperature is higher than the water temperature, the heat stored in the air is transferred to the water by forced convection heat transfer in the heat exchanger to heat the irrigation water in the thermal storage bucket.

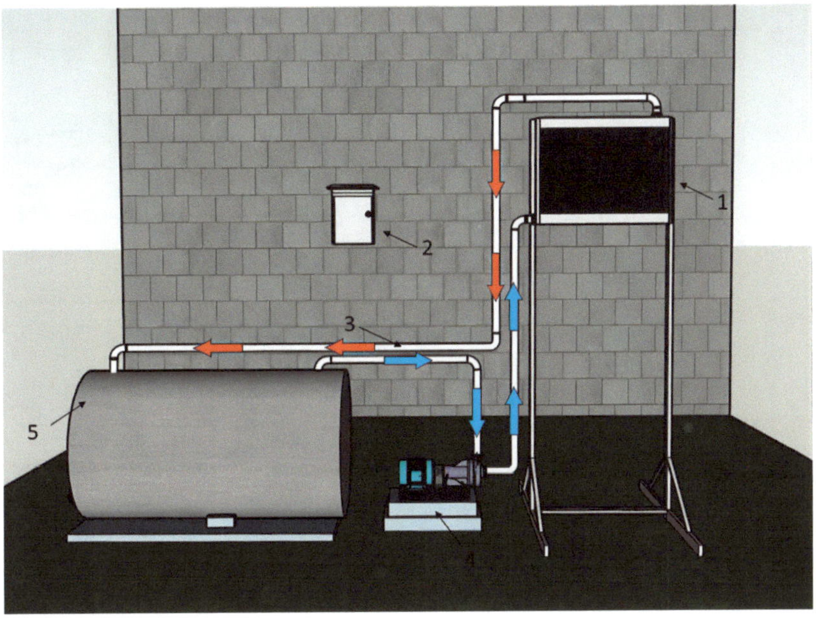

Figure 5. Schematic diagram of the IWHS. **1** denotes the heat exchanger comprising a fan and heat exchanger, **2** denotes the electrical control box, **3** indicates the water circulation pipe, **4** indicates the water pump, and **5** denotes the thermal insulation water storage barrel.

4.2. IWHS Process Parameter Screening

4.2.1. Experimental Greenhouse

The IWHS process parameter screening was conducted from 31 October to 21 December 2022, in the horticulture field of Northwest A&F University, Yangling District, Shaanxi Province (34°17′ N, 108°05′ E). The Chinese solar greenhouse spanned 10.5 m, the east–west length was 15.0 m, and the ridge height was 5.8 m. The rear gravel wall was 3.6 m high.

4.2.2. Experimental Method

- Monitoring of Water Temperature in Solar Water Buckets in the Chinese Solar Greenhouse

The experiment placed a polypropylene polymer (PP) bucket in the middle of the ground of the greenhouse. The bucket volume was 120 L and the bucket contained 100 L of water. The water temperature in the middle of the bucket was determined using a PT100 temperature sensor at 10 min collection intervals. The test monitored the water temperature change in the small solar water bucket (Figure 6). At a maximum indoor air temperature of 26.6 °C and minimum air temperature of 4.4 °C, the water temperature in the solar water bucket increased from the highest 11.6 °C on the first day to 14.7 °C for five consecutive days, and the average water temperature increased by 1.9 °C during the day and decreased by 1.2 °C at night. When the water temperature decreased to 11.7 °C after a cloudy day, it was only 0.1 °C higher than the highest water temperature on the first day, and the increase rate was slow; this was easily affected by low-temperature weather, and it was difficult to reach the appropriate irrigation water temperature. Simultaneously, the small amount of test water resulted in a faster heating rate, while the heating rate in the large solar water bucket was slower, and it was more difficult to meet the required irrigation requirements.

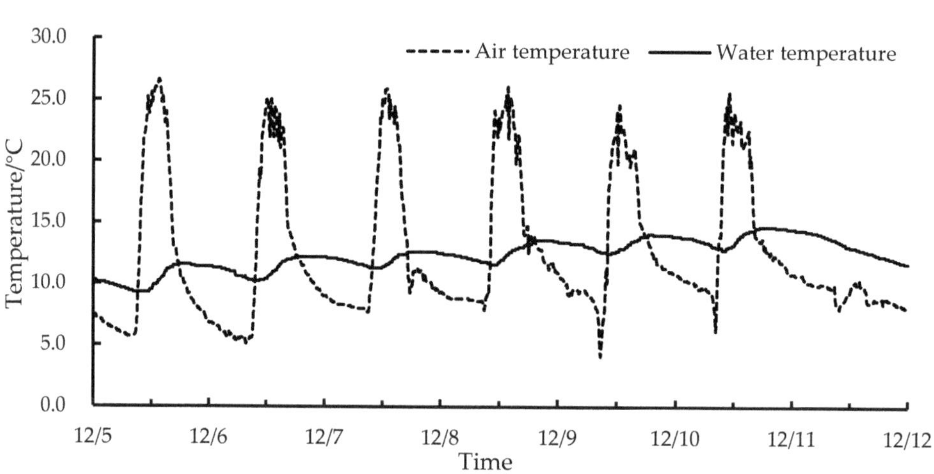

Figure 6. The water temperature changes in the solar water bucket in the solar greenhouse.

- Screening of IWHS Fan Speed

The fan wind speed screening test used three sets of systems to set the wind speeds of 2.3, 3.4, and 4.5 m/s (two repetitions, five consecutive measurements at each time point, and the average value was taken), and other conditions were consistent. The heat exchanger was a harmonica-type, and the overall size was 500 mm × 600 mm × 50 mm. The polypropylene random (PPR) water circulation pipe had a diameter of 32 mm. The pump flow rate was 27 ± 1 L/min. The thermal insulation storage buckets were covered with 10 mm thick thermal insulation cotton with aluminum foil on the exterior.

- Screening of IWHS Heat Exchange Device Height

The air temperature and solar radiation intensity vary at different times and spatial locations within the greenhouse. Therefore, process parameter screening for different heat exchanger heights was conducted. The heat exchanger was positioned at the midpoint of the greenhouse length, 1.5 m from the back wall. The heights were set at 0, 1.5, and 3.0 m above the ground. The water flow rates for all three heights were maintained at 27 ± 1 L/min and with a wind speed of 4.5 m/s. Figure 7 depicts the side view of the experimental measurement points.

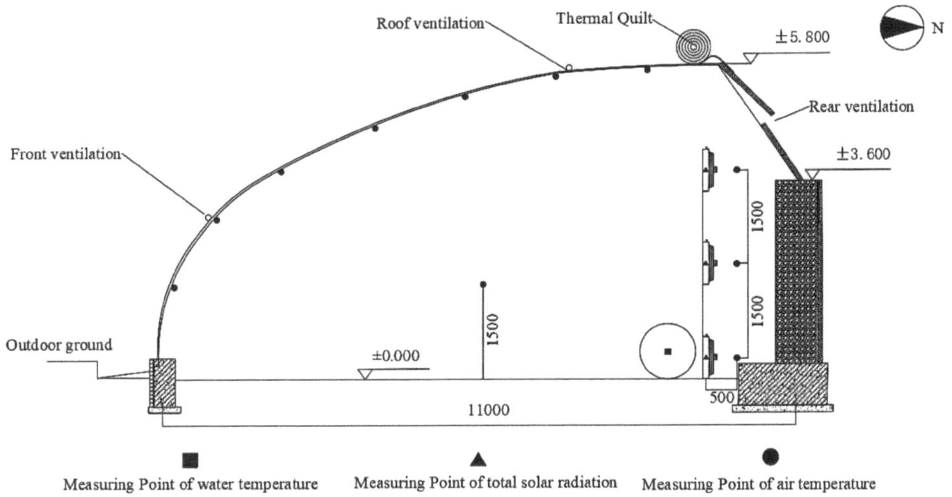

Figure 7. Side view of IWHS performance test measuring points at different heights.

- Screening of IWHS Heat Exchanger Surface Color

The surface color screening test of the heat exchanger involved darkening the outer surface of the device with black paint (experimental group), and the control group was the undarkened surface. The other conditions were the same; the distance between the system and the north wall was 1.2 m, and the influence of the system position was ignored to examine the influence of the device surface darkening on the temperature-increase effect of the system.

4.2.3. Relevant Sensor Testing

The PT100 platinum thermal resistance temperature sensor was used to measure the test greenhouse air temperature (1.5 m above the ground), the water temperature in the middle of each water bucket, and the fan inlet air temperature. The PT100 platinum thermal resistance temperature sensor data were collected using a TCP-518D temperature acquisition module (Gekong Electronics, Beijing, China). The total solar radiation on the heat exchanger surface was measured using the total solar radiation sensor. Water flow was measured using an electronic fuel meter, while the fan operating wind speed was measured using a smart anemometer (Table 6).

4.3. Application Effect Test of IWHS in Lettuce Cultivation

4.3.1. Experimental Greenhouse

The IWHS was used to study the application of warm-water irrigation on lettuce in the Chinese solar greenhouse of Modern Agriculture Innovation Park in Yangling District, Shaanxi Province (34°30′ N, 108°03′ E) from 7 December 2022 to 7 February 2023. The greenhouse runs east–west, with a length of 48 m, clear span of 9.6 m, ridge height of 5.0 m, sinking depth of 0.4 m, and rear wall height of 4.0 m. The wall is made of clay brick masonry, and the front greenhouse roof is covered by thermal insulation at night. During the test, natural ventilation was achieved using bottom- and top-rolled film ventilatio windows. The insulation was lifted at around 08:30 in the morning and dropped at aroun 5:00 pm. Figure 8 depicts the section diagram of the experimental greenhouse.

Table 6. Detailed monitoring instruments information.

Measurement Metrics	Instrument	Model	Measurement Range	Accuracy	Manufacturer
Air and water temperature	PT100 platinum thermal resistance temperature sensor	WZP-GZPT-A	−50~200 °C	0.1 °C	Guizhong Technology, Guizhong, China
Total solar radiation on heat exchanger surface	Total solar radiation sensor	RS-RA-I20-AL	0~1800 W/m^2	1 W/m^2	Jianda Renke, Jinan, China
Circulating pipe water flow	AWT Electronic fuel meter	/	9~100 L/min	±0.5%	Xier Technology, Hangzhou, China
Fan wind speed	Smart anemometer	AS836	0.3~45 m/s	±3%	Wanchuan Electronic, Qingdao, China

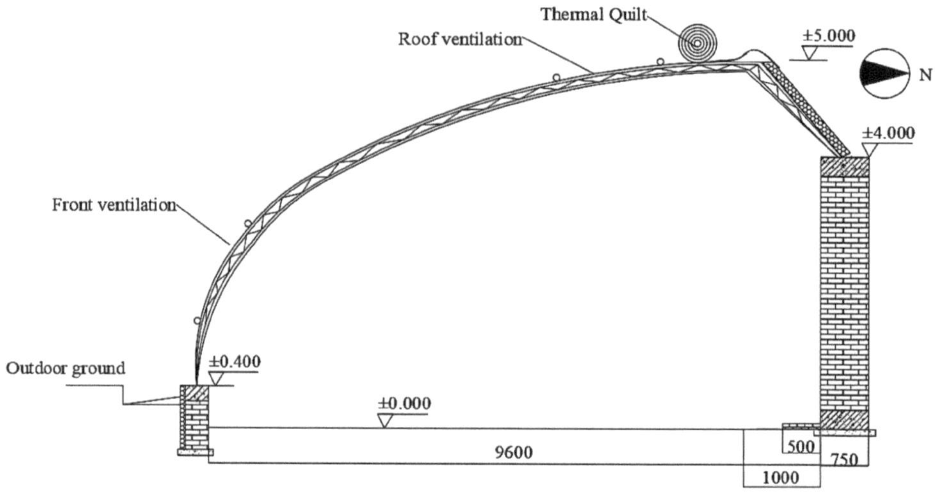

Figure 8. Section diagram of the experimental greenhouse.

4.3.2. Experimental Method

In the system application effect experiment, the test material was loose-leaf lettuc of the variety GRAND RAPIDS TBR. Three-leaf, one-core seedlings of robust growth an uniform size were selected for planting on 7 December 2022. A single cell area was 12.75 m (1.5 m × 8.5 m), and the plants were spaced 30 cm × 30 cm apart (Figure 9). Three cell were established in each T and CK area, and one cell was established at the junction as th quarantine area. Soil cultivation and drip irrigation were used; the irrigation strategy wa artificial irrigation under sunny conditions according to weather conditions. The wate temperature in the experimental group was ≥23 °C, and the management measures wer the same except for the different irrigation water temperatures. According to Li et al. [40 the optimal lettuce root zone temperature is 22~25 °C. The system operation mode was se as follows: the system operated when the air and water temperature controllers detecte that the air temperature and the water temperature of the water storage bucket wer

>26 °C and <23 °C, respectively. The water flow and hot air circulated for rapid heat exchange, and the water temperature continued to rise. The system ceased operating when the temperature controllers detected that the water temperature was >24 °C or the air temperature was <25 °C.

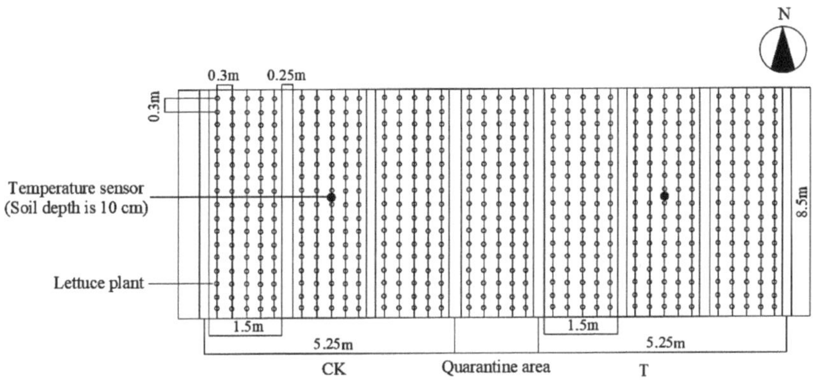

Figure 9. Layout plan of lettuce experimental area.

4.3.3. Relevant Sensor Testing

The PT100 platinum thermal resistance temperature sensor was used to measure the water temperature in the middle of the water body of the bucket, the soil temperature of test area T and control area CK (10 cm depth), and the air temperature of the test greenhouse (1.5 m from the ground). The water flow was measured using the AWT electronic fuel meter, and the fan operating wind speed was measured using the smart anemometer.

The plant height (the distance from the base of the stem to the highest point), and the maximum leaf length and leaf width, of the lettuce were measured with a tape measure, the stem diameter was measured with a vernier caliper, and the number of leaves on each plant was recorded. During the harvest, five whole plants were randomly selected for each treatment, the water from the leaf surface and root system was dried, and the yield of each plant was determined using an electronic balance. The roots from the different treatments were rinsed with deionized water, the roots were scanned using a color image scanner, and the images were stored. The root growth index was analyzed using root analysis software (Table 7).

Table 7. Detailed measuring instruments information.

Instruments/Software	Model	Measurement Range	Accuracy	Manufacturer
Tape measure	GW-580E	0~5 m	0.1 mm	Great Wall Seiko, Ningbo, China
Vernier scale	/	0~150 mm	0.01 mm	Meinaite, Jinhua, China
Electronic balance	BSA224S	10 mg~220 g	0.1 mg	Sartorius, Zhangjiang, China
Color image scanner	J221	/	/	EPSON, Sakata, Japan;
Root analysis software (WinRHIZO Pro 2012a)	STD4800	/	/	Regent, Vancouver, BC, Canada
High precision electronic scale	rz-53	0.1 g~3.0 kg	0.1 g	Daming Technology, Beijing, China

4.4. Relevant Correlation Formula

The system heat collection Q_C was mainly expressed by the heat stored in the water storage tank and calculated as follows:

$$Q_C = \rho_w C_w V \Delta t \tag{1}$$

where ρ_w is the water density in the storage bucket, 1000 kg/m^3; C_w is the specific heat capacity of water, 4183 J/(kg·°C); and V is the water tank volume (m^3). The water quantity in the performance optimization test was 400 L, and the water quantity in the bucket in the lettuce test was 300 L. Δt indicates the change in the initial and end water temperatures of the IWHS.

Q_e represented the overall system energy consumption and was determined as follows:

$$Q_e = (P_{wp} + P_{DF})\Delta t \quad (2)$$

where P_{wp} is the power of the water pump (W), P_{DF} is the fan power (W), and Δt is the system operating time.

Q_s represented the total amount of solar radiation received by the heat exchanger surface (W/m^2) and was calculated as follows:

$$Q_s = \Delta t A I_i \quad (3)$$

where I_i is the instantaneous solar radiation received by the plate surface at time i (W/m^2) and A is the total area of the heat exchanger (m^2), 0.3.

COP is commonly used to evaluate the energy consumption of a heat-collecting system and was derived as follows:

$$COP = \frac{Qc}{Qe} \quad (4)$$

η is the heat-collection ratio, which indicates the ratio of the heat collected by the heat exchanger (the heat stored in the water storage tank) to the solar radiation energy incident on its surface and was calculated as follows:

$$\eta = \frac{Qc}{Qs} \quad (5)$$

5. Conclusions

In this study, the irrigation water-heating system (IWHS) process parameters were designed and screened, and the system application effect on lettuce cultivation was analyzed. The results demonstrated that the average daily water temperature increased by 7.7, 8.[illegible] and 8.6 °C under wind speeds of 2.3, 3.4, and 4.5 m/s, respectively. The system heating capacity was highest at a wind speed of 4.5 m/s. When the heat exchanger was 3.0 m above the ground, the heating capacity increased by 17.8% and 6.0% compared with the change at 0 and 1.5 m, respectively, the operation termination water temperature was 22.0~32.2 °C, and the heating capacity and COP were optimal. Darkening the heat exchanger surface did not affect the system heat-collection capacity. Using IWHS effectively improved the lettuce irrigation water temperature. The system COP was 3.07 in the lettuce greenhouse, and the increased frequency of warm-water irrigation significantly promoted lettuce growth and increased the yield per plant by 15.9%. Further optimization and system improvement are expected to be applied in practical production.

Supplementary Materials: The following supporting information can be downloaded at: https://www.mdpi.com/article/10.3390/plants13050718/s1.

Author Contributions: Conceptualization, Y.C.; Validation, L.G., S.Y. and X.C.; Data curation, L.G. and S.Y.; Formal analysis, L.G.; Methodology, L.G. and X.C.; Resources, Y.C.; Software, S.Y.; Writing—original draft, L.G.; Writing—review and editing, L.G., X.C., S.Y., R.Z. and S.L.; All authors have read and agreed to the published version of the manuscript.

Funding: This research was funded by Shaanxi Province Key R&D Program Project grant number No. 2022ZDLNY03-02, Tibet Autonomous Region Science and Technology Project grant number No. XZ202202YD0002C, and Xinjiang Autonomous Region Key Research and Development Project of grant number No. 2022282251.

Data Availability Statement: The raw data supporting the conclusions of this article will be made available by the authors and included in the Supplementary Material.

Conflicts of Interest: The authors declare no conflicts of interest.

References

1. Maraveas, C.; Karavas, C.S.; Loukatos, D.; Bartzanas, T.; Arvanitis, K.G.; Symeonaki, E. Agricultural greenhouses: Resource management technologies and perspectives for zero greenhouse gas emissions. *Agriculture* **2023**, *13*, 1464. [CrossRef]
2. Li, T.L.; Qi, M.F.; Meng, S.D. Sixty Years of Facility Horticulture Development in China: Achievements and Prospects. *Acta Hortic. Sin.* **2022**, *49*, 2119–2130. [CrossRef]
3. Hu, J.; Fan, G. CFD simulation of heating process of greenhouse irrigation water in wintering period. *Appl. Eng. Agric.* **2017**, *33*, 575–586. [CrossRef]
4. Gao, L.T.; Dai, S.H.; Li, M.; Wang, Z.W.; Zhou, Z.R. Control methods of heating in greenhouse based on crop and energy consumption. *J. Chin. Agric. Mech.* **2021**, *42*, 49–54. [CrossRef]
5. Sakamoto, M.; Suzuki, T. Effect of root-zone temperature on growth and quality of hydroponically grown red leaf lettuce (*Lactuca sativa* L. cv. Red Wave). *Am. J. Plant Sci.* **2015**, *6*, 2350. [CrossRef]
6. Ameen, M.; Wang, X.C.; Yaseen, M.; Umair, M.; Yousaf, K.; Yang, Z.; Soomro, S.A. Performance evaluation of root zone heating system developed with sustainable materials for application in low temperatures. *Sustainability* **2018**, *10*, 4130. [CrossRef]
7. Beyhan, B.; Paksoy, H.; Daşgan, Y. Root zone temperature control with thermal energy storage in phase change materials for soilless greenhouse applications. *Energy Conv. Manag.* **2013**, *74*, 446–453. [CrossRef]
8. He, F.; Hou, Y.; Li, K.; Wei, X.; Liu, Y. Investigation of a root zone heating system for greenhouse seedling and its effects on micro-environment. *Int. J. Agric. Biol. Eng.* **2020**, *13*, 47–52. [CrossRef]
9. He, F.; Tian, J.; Wang, L.; Hou, Y.; Qi, F.; Zhang, Y.; Zhu, L.; Li, Z. Effects of different root zone heating systems on microclimate and crop development in solar greenhouses. *Int. J. Agric. Biol. Eng.* **2022**, *15*, 67–72. [CrossRef]
10. Liu, X.; Yang, S.; Chen, X.; Liu, S.; Zhou, R.; Guo, L.; Sun, Y.; Cao, Y. Performance evaluation of a water-circulating tomato root-zone substrate-cooling system using a chiller and its effect on tomato vegetative growth in Chinese solar greenhouse. *Agronomy* **2022**, *12*, 1922. [CrossRef]
11. Llorach-Massana, P.; Peña, J.; Rieradevall, J.; Montero, J.I. LCA & LCCA of a PCM application to control root zone temperatures of hydroponic crops in comparison with conventional root zone heating systems. *Renew. Energy* **2016**, *85*, 1079–1089. [CrossRef]
12. Miao, Y.M.; Ren, J.R.; Zhang, Y.Z.; Chen, X.C.; Qi, M.Q.; Li, T.L.; Zhang, G.Z.; Liu, Y.L. Effect of low root-zone temperature on photosynthesis, root structure and mineral element absorption of tomato seedlings. *Sci. Hortic.* **2023**, *315*, 111956. [CrossRef]
13. Bi, X.; Wang, X.; Zhang, X. Effects of Different Root Zone Heating Methods on the Growth and Photosynthetic Characteristics of Cucumber. *Horticulturae* **2022**, *8*, 1137. [CrossRef]
14. Tsaballa, A.; Sperdouli, I.; Avramidou, E.V.; Ganopoulos, I.; Koukounaras, A.; Ntinas, G.K. Epigenetic and physiological responses to varying root-zone temperatures in greenhouse rocket. *Genes* **2022**, *13*, 364. [CrossRef]
15. Karnoutsos, P.; Karagiovanidis, M.; Bantis, F.; Chatzistathis, T.; Koukounaras, A.; Ntinas, G.K. Controlled root-zone temperature effect on baby leaf vegetables yield and quality in a floating system under mild and extreme weather conditions. *J. Sci. Food Agric.* **2021**, *101*, 3933–3941. [CrossRef] [PubMed]
16. Yan, Q.; Duan, Z.; Mao, J.; Xun, L.I.; Fei, D. Low root zone temperature limits nutrient effects on cucumber seedling growth and induces adversity physiological response. *J. Integr. Agric.* **2013**, *12*, 1450–1460. [CrossRef]
17. Kawasaki, Y.; Matsuo, S.; Kanayama, Y.; Kanahama, K. Effect of root-zone heating on root growth and activity, nutrient uptake, and fruit yield of tomato at low air temperatures. *J. Jpn. Soc. Hortic. Sci.* **2014**, *83*, 295–301. [CrossRef]
18. He, F.; Thiele, B.; Santhiraraja-Abresch, S.; Watt, M.; Kraska, T.; Ulbrich, A.; Kuhn, A.J. Effects of root temperature on the plant growth and food quality of Chinese broccoli (*Brassica oleracea var. alboglabra Bailey*). *Agronomy* **2020**, *10*, 702. [CrossRef]
19. Jo, W.J.; Shin, J.H. Effect of root-zone heating using positive temperature coefficient film on growth and quality of strawberry (*Fragaria x ananassa*) in greenhouses. *Hortic. Environ. Biotechnol.* **2022**, *63*, 89–100. [CrossRef]
20. Meng, A.J.; Qi, Y.Y.; Fu, Y.B.; Wang, Z.G.; Wang, X.Y.; Feng, Y.Z. Effects of Warming Irrigation on Biomass, Nutrient Uptake and Yield of Cotton. *Xinjiang Agric. Sci.* **2022**, *59*, 558–566.
21. Hendrickson, T.; Dunn, B.L.; Goad, C.; Hu, B.; Singh, H. Effects of elevated water temperature on growth of basil using nutrient film technique. *Hortscience* **2022**, *57*, 925–932. [CrossRef]
22. Wang, L.K.; Huang, J.Y.; Li, P.L.; Weng, J.Y.; Yu, J.; Niu, Q.L. Effects of irrigation water temperature adjustment on the growth of non-heading Chinese cabbage in winter and summer. *Acta Agric. Shanghai* **2022**, *38*, 51–58. [CrossRef]
23. Hooks, T.; Sun, L.; Kong, Y.; Masabni, J.; Niu, G. Effect of Nutrient Solution Cooling in Summer and Heating in Winter on the Performance of Baby Leafy Vegetables in Deep-Water Hydroponic Systems. *Horticulturae* **2022**, *8*, 749. [CrossRef]
24. Abd Elrahman, S.H.; Saudy, H.S.; El Fattah, D.A.A.; Hashem, F.A.E. Effect of irrigation water and organic fertilizer on reducing nitrate accumulation and boosting lettuce productivity. *J. Soil Sci. Plant Nutr.* **2022**, *22*, 2144–2155. [CrossRef]
25. Doklega, S.M.; Saudy, H.S.; El-Sherpiny, M.A.; El-Yazied, A.A.; Abd El-Gawad, H.G.; Ibrahim, M.F.; Abd El-Hady, M.A.; Omar, M.M.; Metwally, A.A. Rhizospheric addition of hydrogel polymer and zeolite plus glutathione mitigate the hazard effects of water deficiency on common bean plants through enhancing the defensive antioxidants. *J. Crop Health* **2024**, *76*, 235–249. [CrossRef]
26. Ramadan, K.M.; El-Beltagi, H.S.; El-Mageed, T.A.A.; Saudy, H.S.; Al-Otaibi, H.H.; Mahmoud, M.A. The Changes in Various Physio-Biochemical Parameters and Yield Traits of Faba Bean Due to Humic Acid Plus 6-Benzylaminopurine Application under Deficit Irrigation. *Agronomy* **2023**, *13*, 1227. [CrossRef]

27. Hu, J.J.; Fan, G.S. Experimental Study on Irrigation Water Heating up Characteristics of Different Heating Facilities in th Greenhouse During Winter. *Water Sav. Irrig.* **2018**, *8*, 103–107.
28. Liu, N.; Wei, Z.Y.; Li, S.; Wang, G.Q.; Wang, Y. Design of comprehensive utilization system of greenhouse's water, fertiliae temperature and gas. *J. Chin. Agric. Mech.* **2018**, *39*, 74–77. [CrossRef]
29. Zhang, J.Z.; Wang, Q.; Wang, Z.H.; Zhen, H.X. Experimental study on irrigation water warming system based on solar phot voltaic/thermal technology. *Trans. Chin. Soc. Agric. Eng.* **2021**, *37*, 72–79. [CrossRef]
30. Bao, E.C.; Cao, Y.F.; Zou, Z.R.; Shen, T.T.; Zhang, Y. Research progress of thermal storage technology in energy-saving sola greenhouse. *Trans. Chin. Soc. Agric. Eng.* **2018**, *34*, 1–14. [CrossRef]
31. Xia, T.; Li, Y.; Sun, Z.; Wan, X.; Sun, D.; Wang, L.; Liu, X.; Li, T. Performance study of an active solar water curtain heating syste for Chinese solar greenhouse heating in high latitudes regions. *Appl. Energ.* **2023**, *332*, 120548. [CrossRef]
32. Xu, W.; Guo, H.; Ma, C. An active solar water wall for passive solar greenhouse heating. *Appl. Energ.* **2022**, *308*, 118270. [CrossRe
33. Fang, H.; Yang, Q.C.; Liang, H.; Wang, S. Experiment of temperature rising effect by heat release and storage with shallow wate in solar greenhouse. *Trans. Chin. Soc. Agric. Eng.* **2011**, *27*, 258–263. [CrossRef]
34. Sun, W.T.; Guo, W.Z.; Xu, F.; Wang, L.C.; Xue, X.Z.; Li, Y.K.; Chen, Y.P. Application effect of surplus air heat pump heating syste in Chinese solar greenhouse. *Trans. Chin. Soc. Agric. Eng.* **2015**, *31*, 235–243. [CrossRef]
35. Yue, H.F.; Zhu, N.; Hu, X.Y.; Zhang, X.M.; Meng, F.Y.; An, S.W.; LI, L.F.; Chang, Y. Effects of Heating Irrigation on Growth Yiel and Quality of Strawberry. *Vegetables* **2023**, *9*, 24–28.
36. Zhang, Y.; Feng, X.L.; Zhao, S.M.; Wang, Q.R.; Ren, X.M. Effects of Air Circulator on Environmental Parameter and Toma Growth in Solar Greenhouse. *China Veg.* **2016**, *9*, 52–57.
37. Song, W.T.; Li, H.; Wang, P.Z.; Wang, X.Z.; Shao, Q.X.; He, X.Y.; Li, M.; Cheng, J.Y.; Meng, L.Q. Design and application effe of surface cooler-fan heat collection and release system—Taking Ningcheng large-span external insulation greenhouse as a example (Part 2). *Agric. Eng. Technol.* **2020**, *40*, 42–48. [CrossRef]
38. Deng, H.L.; Tian, J.C.; Ou, Y.Z. Effect of Irrigation Water Temperature on Soil Temperature in Vegetable Root Zone in Greenhous *Ningxia Eng. Technol.* **2018**, *17*, 97–101.
39. Yamori, N.; Levine, C.P.; Mattson, N.S.; Yamori, W. Optimum root zone temperature of photosynthesis and plant growth depenc on air temperature in lettuce plants. *Plant Mol. Biol.* **2022**, *110*, 385–395. [CrossRef]
40. Li, R.R.; Zhu, Y.L.; Takagaki, M.; Yamori, W.; Yang, L.F. Effects of root temperature on the growth and mineral elements content hydroponically-grown lettuce. *Acta Agric. Shanghai* **2015**, *31*, 48–52. [CrossRef]

plants

MDPI

Article

Serratia marcescens LYGN1 Reforms the Rhizosphere Microbial Community and Promotes Cucumber and Pepper Growth in Plug Seedling Cultivation

Xu Zhang [1,†], Jinxin Peng [1,†], Xiaodong Hao [1], Guifang Feng [1], Yanhui Shen [2], Guanghui Wang [2] and Zhiqun Chen [1,*]

1 College of Life Science, Linyi University, Linyi 276000, China; zhangxu1787@163.com (X.Z.); 13263127966@163.com (J.P.); haoxiaodong@lyu.edu.cn (X.H.); jyxtw@126.com (G.F.)
2 Shandong (Linyi) Institute of Modern Agriculture, Zhejiang University, Linyi 276000, China; shenyh2019@126.com (Y.S.); wgh17806288826@163.com (G.W.)
* Correspondence: chenzhiqun@lyu.edu.cn
† These authors contributed equally to this work.

Abstract: The vegetable plug seedling plays an important role in improving vegetable production. The process of plug seedling contributes to high-quality vegetable seedlings. The substrate composition and chemical fertilizer are widely studied to promote seedling growth. However, little is known about the effect of beneficial bacteria in the rhizosphere microbial community and vegetables' growth during plug seedling. The use of beneficial microbes to promote vegetable seedling growth is of great potential. In this study, we showed that the *Serratia marcescens* strain LYGN1 enhanced the growth of cucumber and pepper seedlings in plug seedling cultivation. The treatment with LYGN1 significantly increased the biomass and the growth-related index of cucumber and pepper, improving the seedling quality index. Specifically, LYGN1 also improved the cucumber and pepper root system architecture and increased the root diameter. We applied high-throughput sequencing to analyze the microbial community of the seedlings' rhizosphere, which showed LYGN1 to significantly change the composition and structure of the cucumber and pepper rhizosphere microbial communities. The correlation analysis showed that the Abditibacteriota and Bdellovibrionota had positive effects on seedling growth. The findings of this study provide evidence for the effects of *Serratia marcescens* LYGN1 on the cucumber and pepper rhizosphere microbial communities, which also promoted seedling quality in plug seedling cultivation.

Keywords: *Serratia marcescens*; cucumber; pepper; microbial community; seedlings growth

Citation: Zhang, X.; Peng, J.; Hao, X.; Feng, G.; Shen, Y.; Wang, G.; Chen, Z. *Serratia marcescens* LYGN1 Reforms the Rhizosphere Microbial Community and Promotes Cucumber and Pepper Growth in Plug Seedling Cultivation. *Plants* **2024**, *13*, 592. https://doi.org/10.3390/plants13050592

Academic Editors: Weijie Jiang and Wenna Zhang

Received: 14 December 2023
Revised: 13 February 2024
Accepted: 19 February 2024
Published: 22 February 2024

1. Introduction

Vegetable plug seedling could significantly improve seedling quality, having become the widely used way to produce vegetable seedlings [1,2]. Plant growth is often limited by diverse biotic and abiotic stress [3]. For dealing with these environmental pressures in vegetable cultivation, the improvement of plant seedlings quality becomes more and more important. According to the statistics, China produces up to 350 billion plants of professional plug seedlings annually [4], and approximately 60% of the world's vegetable varieties currently use plug seedling technology [2]. Thus, due to the fact that the demands for vegetable yield and quality are getting higher, researchers are now seeking effective methods to enhance seedling quality.

Cucumber (*Cucumis sativus* L.) is a worldwide cultivated crop which is affected by numerous factors, including temperature, relative humidity, irrigation, fertigation, and disease incidence [5]. Therefore, cucumber plug seedlings are widely used in the improvement of cucumber quality and yield [6]. The plug seedling cultivation of pepper

(*Capsicum annuum* L.) also plays a vital role in improving pepper growth and yield after b ing transformed into soil cultivation [1]. The plant growth-promoting rhizobacteria (PGP showed beneficial effects on plant growth [7,8] and also contributed to plant respons under different environmental stress [9–11]. They can promote plant growth, increas plant biomass, maintain soil fertility, and protect the plant from phytopathogens. PGP have been increasingly used in vegetable seedling cultivation, proving to improve see ing growth and promote disease inhibition [12]. There still is knowledge gap about th consequences of the application of PGPR on the rhizosphere microbial community an seedling growth in plug seedling cultivation. *Serratia marcescens* is a genetically divers species which include rhizobacteria with high potential to alleviate biotic stress [13]. Th strain of *Serratia marcescens* in this paper was isolated from the cucumber rhizosphere c plants which were grown in a cucumber–amaranth intercropping system [14]. In previou studies, the *Serratia marcescens* NBRI1213 and S-JS1 were respectively observed to promo plant growth and alleviate biotic stress in betelvine [15] and rice [16]. The genomic stud of the *Serratia marcescens* strain showed the potential to be used in plant growth promotio or as bioinoculants in agricultural cultivation [17]. Several reports have shown that the a plication of PGPR, including *Serratia* sp., promotes plant growth and yield of economical important crops [18,19], suggesting the potential of bacteria of this species to improve plar growth in seedling production.

Specific PGPR showed beneficial effects on plant growth, but there is still very litt research conducted in plug seedlings cultivation using artificial cultivation substrates. Targe ing the requirements for high-quality seedlings in cucumber and pepper cultivation, speci attention should be paid to the application of PGPR, which play critical roles in plant growt This study was conducted to evaluate the effects of LYGN1 on seedling quality, as well as o the cucumber and pepper rhizosphere microbial community in plug seedling cultivation.

2. Materials and Methods

2.1. Plant Material and Bacterial Strain

Cucumber (*Cucumis sativus* L. cv. 'Zhongnong 62') and pepper (*Capsicum annuum* cv. 'Shifeng 802') were used in the experiment. All seeds were sterilized using 70% ethanc and 3% sodium hypochlorite solution and rinsed with sterile water [20]. *Serratia marcescen* LYGN1 was isolated from the rhizosphere of cucumber in the cucumber–amaranth inte cropping system. The strain was deposited at the China General Microbiological Cultur Collection Center (CGMCC) under the accession number CGMCC No. 27660.

2.2. Experimental Design

The experiment was carried out in a controlled-environment greenhouse at the exper mental station of Linyi University, located in the city of Linyi, Shandong Province, Chin The cucumbers and peppers were planted in a substrate consisting of peat, vermiculit and perlite at a ratio of 2:1:1. A single colony was selected in an LB liquid culture mediur and incubated overnight at 37 °C to obtain a bacterial suspension. The absorbance of th bacterial suspensions was measured by a spectrophotometer and adjusted to 1 (OD_{600}) wit a sterile LB medium. The germinated cucumber and pepper seeds were sown in 50-hol trays, which are specifically used for seedling cultivation.

The seedlings of cucumber and pepper were treated with the bacterial suspensio of LYGN1 when the cotyledon fully expanded. The bacterial suspension was applied t cucumber and pepper by irrigating it into the root-zone area, with every seedling bein irrigated with 20 mL of bacterial suspension [21].

2.3. Sampling and Determination of Seedlings Growth

Two time points during the cucumber and pepper seedlings growth were selecte For cucumber, we sampled the cucumber seedlings at 20 and 25 days after inoculatio with bacterial suspensions and 44 and 53 days for pepper seedlings. For every treatmen six individual seedlings were sampled to measure the different index. First, the plan height was measured using measuring tape, and the stem thickness was determined by measuring the diameter of the stem base. The total leaf area of peppers was scanned by a scanner (EPSON V800, Seiko Epson Inc., NKS, Tokyo, Japan) and analyzed using software WinRHIZO (LC4800-II LA2400, Sainte Foy, QC, Canada). The cucumber total leaf area was approximated by the product of the measured leaf length, leaf width, and correction factor [22]. The seedlings' biomass was divided into the shoot biomass and root biomass, which were separately oven-dried at 80 °C until constant weight. The growth rate of cucumber and pepper was calculated by the dividing the change in biomass weight by the number of days for each interval for both cucumber and pepper. Root–shoot ratios were determined by dividing root biomass by shoot biomass. The SPAD chlorophyll values of seedling leaves were measured using a chlorophyll meter, SPAD-502 plus (Konica Minolta, Tokyo, Japan). The seedling vigor index of cucumbers and peppers was calculated according to previous research [23], as follows:

$$\text{Seedlings vigor index} = \text{TDW}/(\text{PH}/\text{SD} + \text{SDW}/\text{RDW})$$

where TDW, SDW and RDW present total dry weight (g), shoot dry weight (g) and root dry weight (g), respectively; and PH and SD are plant height (cm) and stem diameter (mm), respectively.

2.4. Root System Architecture Analysis

For the cucumber and pepper root architecture analysis, the cucumber and pepper seedlings at two time points at the end of the experimental trial were carefully extracted from the substrate, scanned using a scanner (EPSONV 800), and analyzed by software WinRHIZO. For each species, the following root architecture measurements were measured: total root length, the number of root tips, average root diameter, and the total root surface area.

2.5. Microbial Community of Cucumber and Pepper Rhizosphere

The rhizosphere microbial community genomic DNA of cucumber and pepper was extracted and purified from about 400 mg substrate using the PowerSoil DNA isolation kit (QIAGEN Inc., Santa Clarita, CA, USA). The primers 515F (5′-GTGCCAGCMGCCGCGGTAA-3′) and the reverse primer 806R (5′-GGACTACHVGGGTWTCTAAT-3′) [17,24], as well as the primers ITS1F (5′-CTTGGTCATTTAGAGGAAGTAA-3′) and the reverse primer ITS2R (5′-GCTGCGTTCTTCATCGATGC-3′) were used to amplify the 292 bp fragment of the V4 region in the 16S rRNA gene and for the amplification of the 300 bp fragment of the fungal ITS gene [18,25], respectively. PCR was performed under the following conditions: 95 °C for 3 min, followed by 27 cycles at 95 °C for 30 s, 55 °C for 30 s and 72 °C for 45 s, with a final extension at 72 °C for 10 min. The sample libraries for sequencing were prepared according to the MiSeq Reagent Kit Preparation Guide (Illumina, Inc., San Diego, CA, USA) and sequenced using the Illumina MiSeq platform. The high-throughput sequencing was performed by Majorbio (Shanghai Majorbio Bio-pharm Technology Co., Ltd., Shanghai, China).

Illumina sequences of 16S and the ITS were processed and sequentially quality-filtered using Fastp (version 0.19.6) [26]. Pair-end reads were merged with a minimum overlap using Flash (1.2.11) [27]. After removing chimeric sequences, the remaining sequences were binned into operational taxonomic unit (OTU) with 97% similarity and the representative sequence for each OTU was taxonomically classified via the Ribosomal Database Project's classifier [28], the SILVA database (version 138) [29] for bacteria, and the UNITE (version 8.0) for fungi [30]. All OTUs identified as belonging to chloroplast and mitochondria were removed from the data set. Then, the representative sequences for each OTU were aligned using PyNAST [31] in QIIME [32] and carried out by Uparse software (version 11) [33].

2.6. Statistical Analysis

The data of seedlings growth were subjected to analysis using Student' *t*-test, carried out by SPSS 26.0. To avoid the bias caused by different sequencing depths, the least number of obtained sequences from all microbial samples was used for normalization. The Shannon index, Chao1, and the Simpson index were applied to directly compare the α-diversity the root-zone soil microbial community. Non-metric multidimensional scaling was used assess the β-diversity of the microbial community. The rhizosphere microbial communi data were analyzed using the online platform Majorbio Cloud Platform (www.majorbi com, accessed on 19 January 2023). For correlation analysis, Pearson product–mome correlation coefficient (Pearson's r) analysis was performed. The raw sequencing data of th bacteria and fungi were submitted into the NCBI Sequence Read Archive (SRA) databa (Accession Number: PRJNA1051178).

3. Results

3.1. The Cucumber and Pepper Seedling Traits and Root Morphology

The application of LYGN1 had the same trends in the biomass accumulation of c cumber and pepper. The shoot and root biomass of cucumber were significantly increase after being inoculated with the bacterial suspensions ($p < 0.001$) (Figure 1). In this stud the shoot and root were sampled at two timepoints following the cucumber and pepp growth. Positive effects of LYGN1 application on seedling biomass accumulation we observed at two sampled timepoints (Figure 1). As showed in the representative pictu of cucumber and pepper seedlings at 25 and 53 DAI, respectively, the seedlings treate with LYGN1 were obviously larger than the control (CK) treatments (Figure 1a,b). We als measured several growth indexes in cucumber and pepper plants to evaluate the effects LYGN1 application. Through many plant traits, we demonstrated that the effect of pla growth promotion was strongly induced in the LYGN1 treatment group (Figure 2). Th plant height and stem diameter were both significantly improved under LYGN1 treatme which showed a great promoting effect. Meanwhile, the growth rate of cucumber an pepper seedlings was significantly higher in the LYGN1-treated plants, which also ha the higher chlorophyll fluorescence (SPAD). Taken together, LYGN1 treatment enhance biomass accumulation and promoted seedling growth.

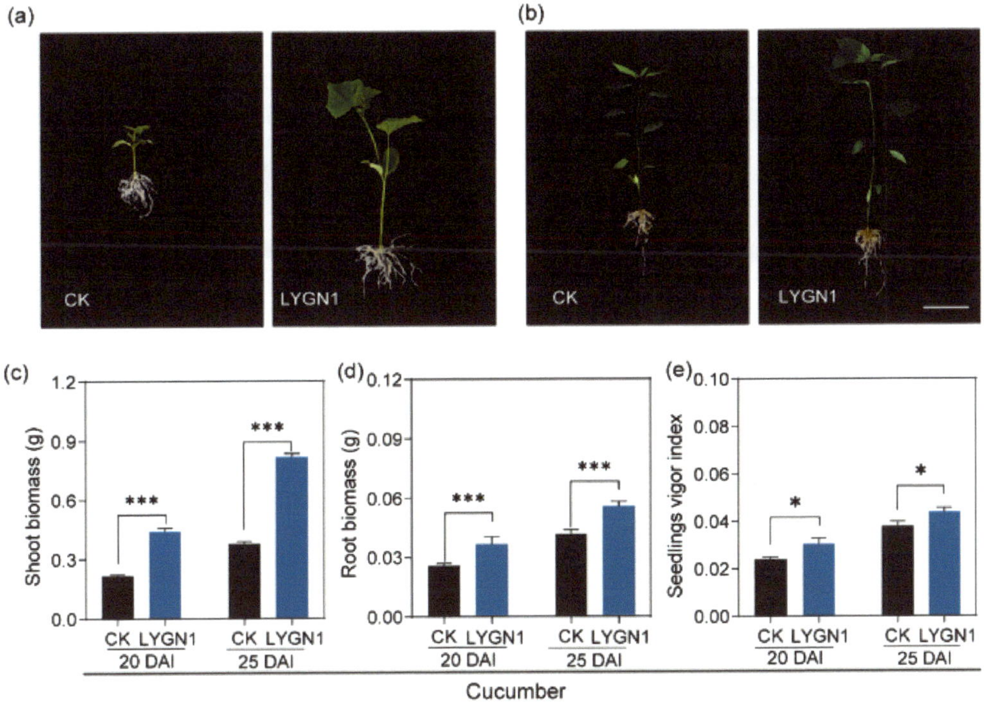

Figure 1. *Cont.*

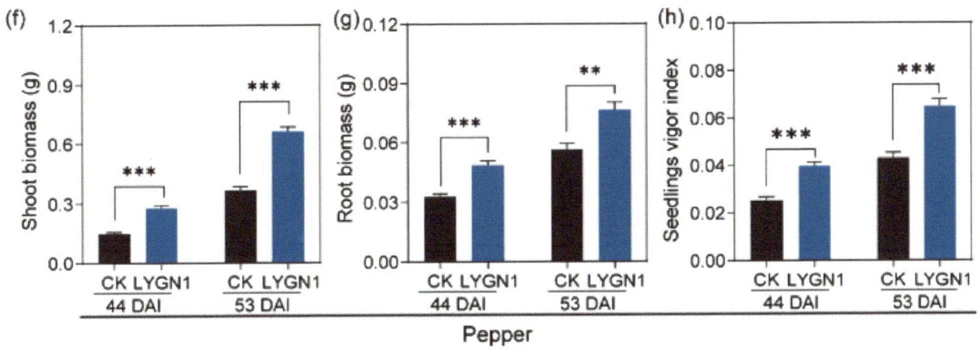

Figure 1. Effects of LYGN1 treatment on cucumber and pepper biomass accumulation and seedlings vigor index. CK: control; LYGN1: cucumber and pepper seedlings treated with *Serratia marcescens* LYGN1; DAI: days after inoculation. Representative pictures of cucumber (**a**) and pepper (**b**) seedlings treated with control and LYGN1, scale bar: 5 cm. The biomass of shoot (**c**) and root (**d**), and the seedlings vigor index (**e**) in cucumber seedlings. The biomass of shoot (**f**) and root (**g**), and the seedlings vigor index (**h**) in pepper seedlings. Data are shown as mean ± SEM, *p*-values, calculated using Student' *t*-test, * $p < 0.05$, ** $p < 0.01$, *** $p < 0.001$, $n = 6$.

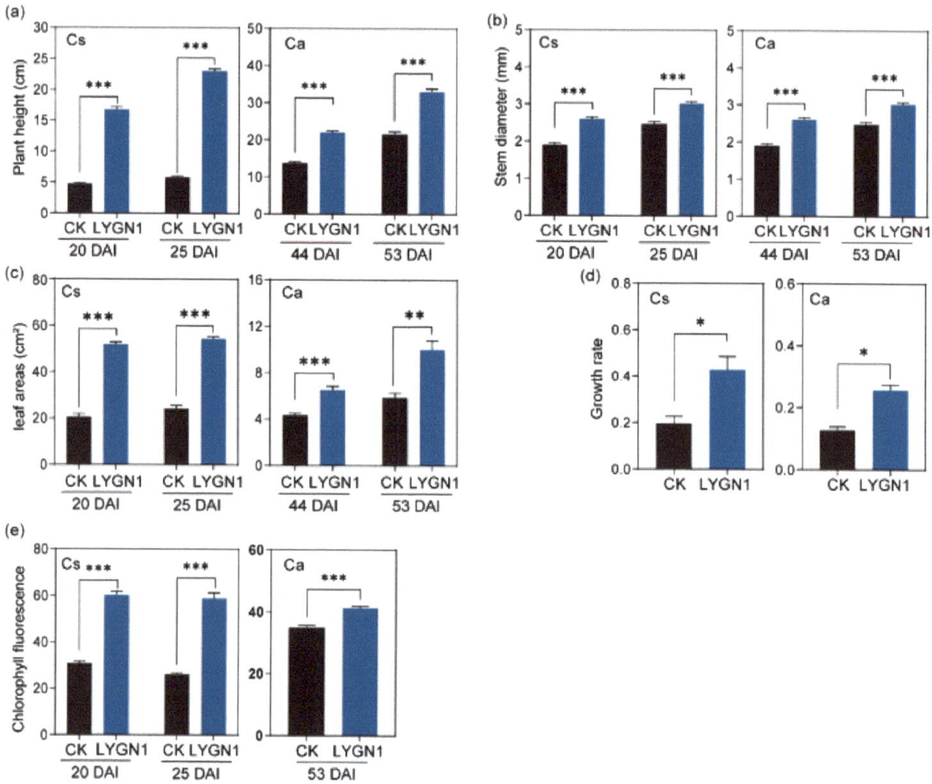

Figure 2. Effects of LYGN1 treatment on growth-related indexes of cucumber and pepper seedling CK: control; LYGN1: cucumber and pepper seedlings treated with *Serratia marcescens* LYGN1; DA days after inoculation; Cs: *Cucumis sativus;* Ca: *Capsicum annuum.* (**a**) The plant height of cucumbe and pepper seedlings at different sampled timepoints. (**b**) The stem diameter of cucumber and peppe seedlings at different sampled timepoints. (**c**) The leaf area of cucumber and pepper seedlings a different sampled timepoints. (**d**) The growth rate of cucumber and pepper seedlings at differen sampled timepoints. (**e**) The Chlorophyll fluorescence of cucumber and pepper seedlings at differen sampled timepoints. Data are shown as mean ± SEM, *p*-values, calculated using Student' *t*-tes * $p < 0.05$, ** $p < 0.01$, *** $p < 0.001$, $n = 6$.

All the parameters measured and calculated by WinRHIZO system software are show in Figure 3. The root system of cucumber (Cs) and pepper (Ca) seedlings was larger and ha more fine roots (Figure 3a,b). Especially for the pepper seedlings, the total root length was sig nificantly higher in the LYGN1 treatment (S) at both sampling timepoints (Figure 3c). Anothe important component of a functional root system is the root surface area, which represents th total area of the root system that is in contact with the substrate. LYGN1 treatment significantl increased root surface, especially for the pepper root at 44 DAI (Figure 3d). In this stud LYGN1 could also affect root volume and root tip development. Overall, both cucumber an pepper seedlings showed a significant promoting effect (Figure 3f). Based on the diamete of the root, roots were categorized in five grades (Figure 4). LYGN1 application facilitate the growth of many more lateral roots, with thicker diameter, in the cucumber and peppe seedlings (Figure 4). Especially during the growth of pepper seedlings, this phenomenon wa more significant (Figure 4b). In short, the application of LYGN1 significantly promoted the growth of cucumber and pepper seedlings.

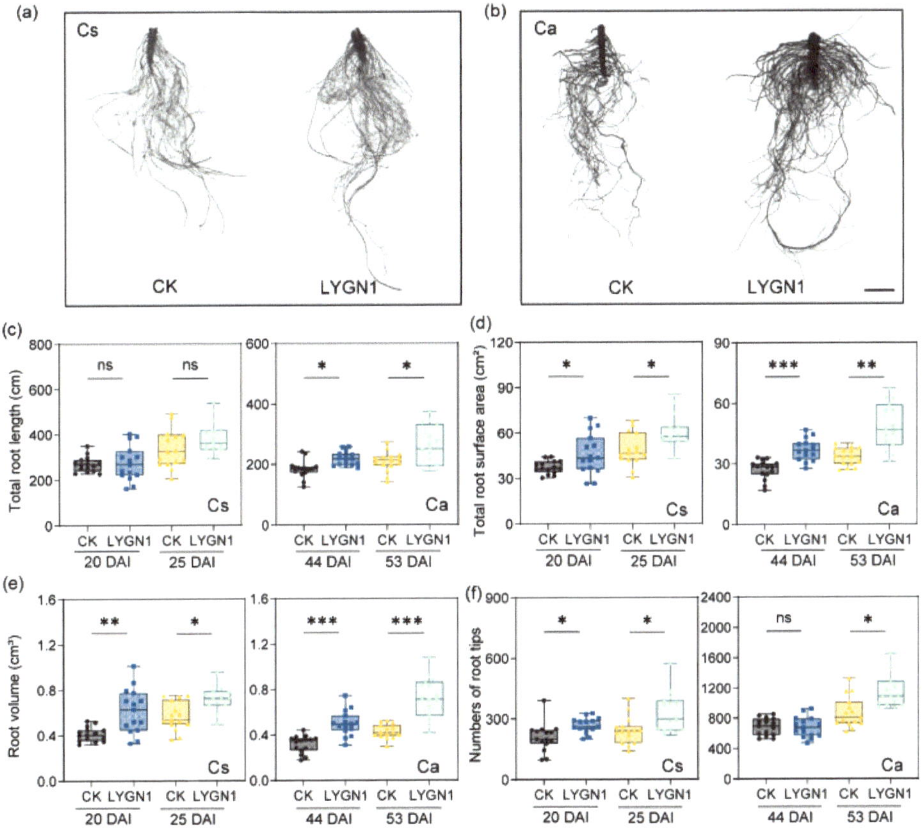

Figure 3. Effects of LYGN1 treatment on root system growth of cucumber and pepper seedlings. CK: control; LYGN1: cucumber and pepper seedlings treated with *Serratia marcescens* LYGN1; DAI: days after inoculation; Cs: *Cucumis sativus*; Ca: *Capsicum annuum*. Representative pictures of cucumber (**a**) and pepper (**b**) root system treated with control and LYGN1, scale bar: 5 cm. (**c**) The total root length of cucumber and pepper seedlings. (**d**) The total root surface area of cucumber and pepper seedlings. (**e**) The root volume of cucumber and pepper seedlings. (**f**) The number of root tips of cucumber and pepper seedlings. The p-values were calculated using Student' t-test, * $p < 0.05$, ** $p < 0.01$, *** $p < 0.001$, ns: no significance, $n = 15$.

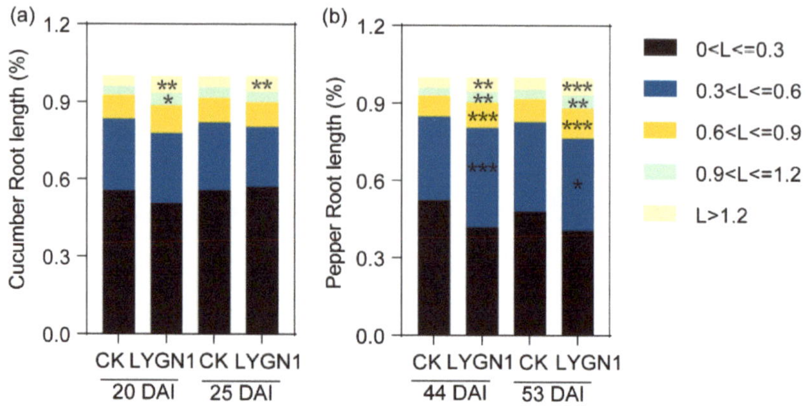

Figure 4. Effects of LYGN1 treatment on root system architecture. CK: control; LYGN1: cucumb and pepper seedlings treated with *Serratia marcescens* LYGN1; DAI: days after inoculation. The roo were divided into five groups, according to their diameters (L, mm). (**a**) The root classification cucumber seedlings' root system. (**b**) The root classification of the pepper root system. The *p*-valu between the control and LYGN1 treatment were calculated using Student' *t*-test, * $p < 0.05$, ** $p < 0.0$ *** $p < 0.001$, $n = 15$.

3.2. Microbial Community Diversity of Cucumber and Pepper Rhizosphere

The rhizosphere microbial community is known to greatly promote plant growt We also investigated the effects of the application of LYGN1 on cucumber and peppe rhizosphere microbial community. The smallest numbers of bacterial sequences (33,820 s quences per sample) and fungi (31,098 sequences per sample) were normalized to avoid th deviation caused by the effects of different sequencing depths. A total of 2,776,766 raw read were obtained for all the samples described in this study. After filtering, 2,224,310 read and 4900 OTUs were obtained. The Shannon index of cucumber at the OTU level wa higher after the LYGN1 application compared to the control. Both bacterial and fung communities showed this trend in cucumber. However, the application of LYGN1 slightl decreased the Shannon index of fungal communities in pepper seedlings (Figure 5). Thes results showed the different response of LYGN1 application in cucumber and pepper. I the bacterial community, the relative abundance of Proteobacteria was significantly highe in the LYGN1 treatment in both cucumber and pepper rhizospheres (Figure 6a). Esp cially in the pepper rhizosphere, the relative abundance of *Proteobacteria* increased 16.29 in the LYGN1 treatment compared to the control after 53 DAI. The relative abundanc of Bacteroidota in the rhizosphere of cucumber and pepper showed an opposite tren with an increasing trend in pepper (Figure 6a,b). The addition of LYGN1 had a great impact on Ascomycota in the fungal community, reducing its relative abundance in th rhizosphere of cucumber and pepper (Figure 6c,d). These results showed that *Serrat marcescens* may affect the colonization of Ascomycota on the cucumber and pepper rhiz spheres. The Venn diagrams show the shared and unique OTUs in the bacterial and fung community (Figure 6e,g), which exhibited different effects on the cucumber and peppe microbial communities. The principal coordinates analysis (PCoA) of the bacterial an fungal communities also showed the significant separation between LYGN1 treatment an control conditions in both cucumber and pepper rhizospheres (Figure 7), demonstratin the effect of LYGN1 application on the microbial communities of cucumber and peppe seedling rhizospheres.

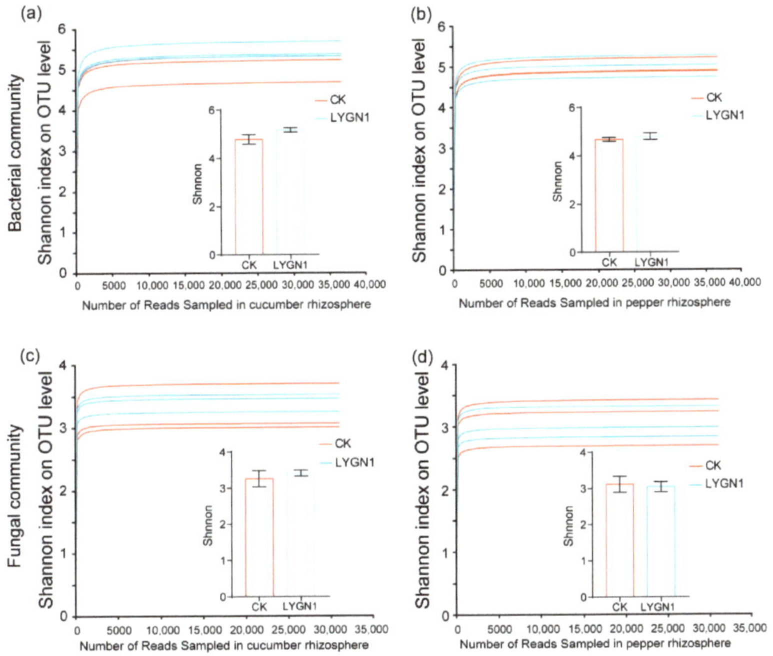

Figure 5. Effects of LYGN1 treatment on α-diversity of cucumber and pepper rhizosphere microbial community. CK: control; LYGN1: cucumber and pepper seedlings treated with *Serratia marcescens* LYGN1; OUT: operational taxonomic unit. The Shannon index of the bacterial community in cucumber (**a**) and pepper (**b**) under LYGN1 treatment. The Shannon index of the fungal community in cucumber (**c**) and pepper (**d**) under LYGN1 treatment. Data are shown as mean ± SEM, $n = 3$.

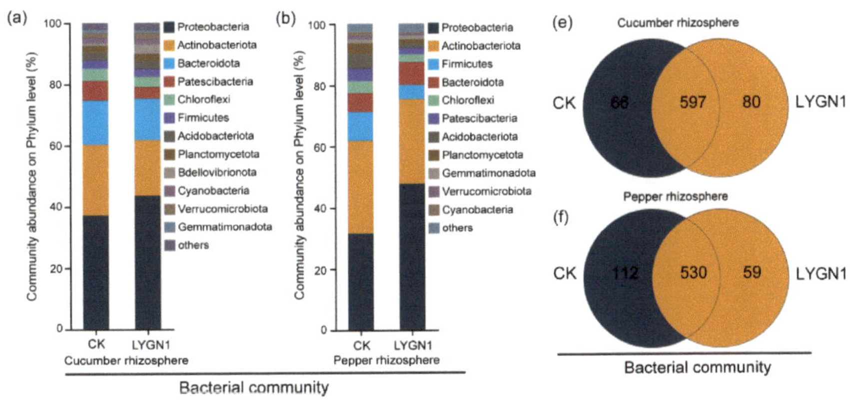

Figure 6. *Cont.*

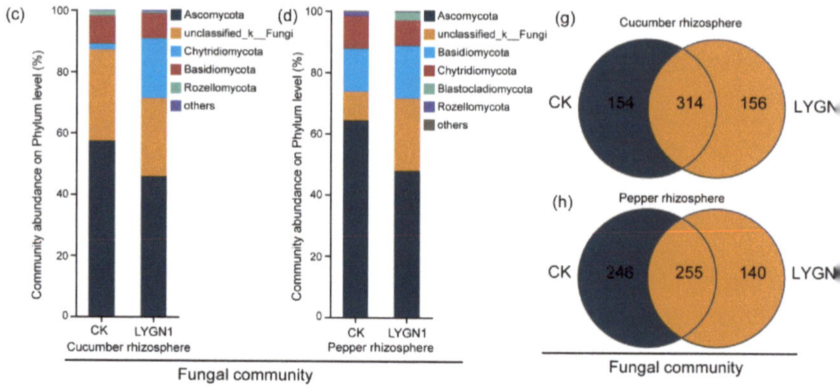

Figure 6. Effects of LYGN1 treatment on the composition of cucumber and pepper rhizosphe microbial communities. CK: control; LYGN1: cucumber and pepper seedlings treated with *Serrat marcescens* LYGN1. The relative abundance of cucumber (**a**) and pepper (**b**) rhizosphere bacteri communities at phylum level. The Venn diagram shows the numbers of specific and shared OTU between the cucumber (**e**) and pepper (**f**) rhizospheres under LYGN1 treatments. The relativ abundance of cucumber (**c**) and pepper (**d**) rhizosphere microbial communities at phylum level. Th Venn diagram shows the numbers of specific and shared OTUs between cucumber (**g**) and pepp (**h**) rhizosphere microbial communities after LYGN1 treatment.

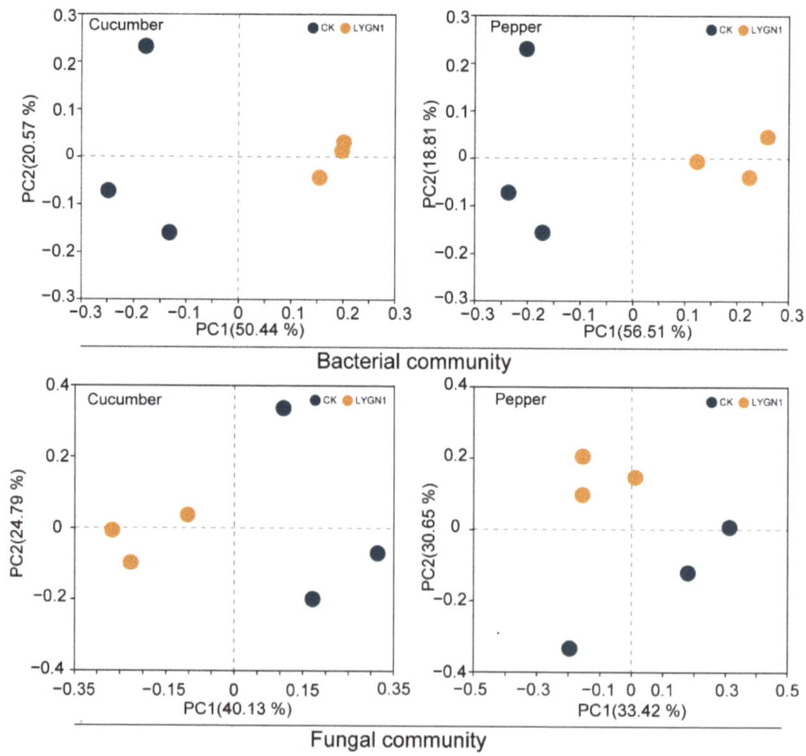

Figure 7. Effects of LYGN1 treatment on the β-diversity of cucumber and pepper rhizospher microbial communities. Principal coordinate analysis (PCoA) was based on Bray–Curtis distance CK: control; LYGN1: cucumber and pepper seedlings treated with *Serratia marcescens* LYGN1.

3.3. Correlation Analysis of Phylum-Level Microbial Groups and Seedlings Index

We measured the different plant growth and root system index of cucumber and pepper seedlings, and the correlation analysis revealed the association between the seedlings and annotated microbial community at phylum level (Figure 8). The *Abditibacteriota* significantly affect the cucumber growth positively, especially for the shoot biomass and seedlings growth rate (Figure 8) but affect pepper growth slightly. The relative abundance of the phylum *Chloroflexi*, *Cyanobacteria*, *Deinococcota*, and *Dependentiae* was significantly decreased in cucumber rhizosphere but showed negative effects on cucumber shoot and root growth. The changes in fungal community caused different effects on the cucumber and pepper indicators (Figure 8). Pearson's correlation analysis showed significant positive correlations between *Ascomycota* and seedling indicators, but in the cucumber and pepper rhizosphere, the phylum *Chytridiomycota* and *Mortierellomycota* showed opposite correlations with the different indictors. All these results showed the effects of the LYGN1 application on microbial communities and the subsequent effects on cucumber and pepper seedling growth.

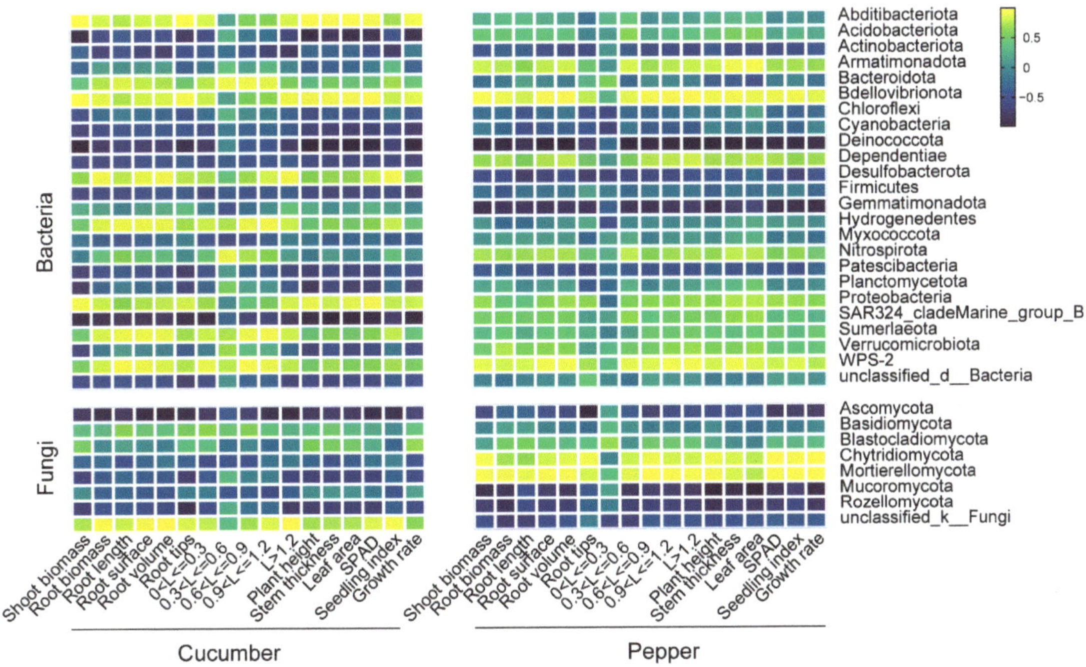

Figure 8. Correlation analysis between the phylum-level microbial groups and the seedling growth index. The heatmap shows the positive and negative correlations between the changes in phylum-level microbial groups and different seedlings index. Pearson's product–moment correlation coefficient (Pearson's r) analysis was performed. The data used for the analysis of all samples were taken from the latest sampling period. The seedlings growth-related index includes shoot and root biomass, stem thickness, leaf area, SPAD (chlorophyll fluorescence), seedling index, and growth rate. The root system-related index includes root length, root surface area, root volume, and root tips, as well as the five group of roots classified according to their diameters (L, mm).

4. Discussion

Previous studies showed that the application of PGPR could promote plant growth by reforming the rhizosphere microbial environment [6,34]. In this study, we showed that LYGN1 treatment has the same trends. The structure of the bacterial and fungal communities in both cucumber and pepper rhizosphere was reformed. Meanwhile, the application of LYGN1 in cucumber and pepper plug seedlings showed significant promotion effe on both seedling growth and root development. This promotion effect showed the san trends in the two sampled timepoints after the LYGN1 treatments.

The compact seedlings were demonstrated to lead to healthier plants, higher yielc and improved nutritional qualities compared to weak seedlings after transplanting [3. Therefore, the cultivation of high-quality vegetable seedlings is vital for increasing crc yield and farmer income. In our study, the seedling vigor index was significantly higher the LYGN1 treatment, showing its potential in cucumber and pepper seedling cultivatic Plant growth-promoting rhizobacteria (PGPR) can promote biomass accumulation ar plant growth [36]. This finding is the same as that in our study, which demonstrated th the application of LYGN1 could significantly promote the growth of cucumber and pepp Studies showed that *S. marcescens* can promote the growth of various plants by producir IAA [17,19,37], which can also significantly increase root growth. The chlorophyll fluore cence in this study had a significant increase after the LYGN1 application, so the enhance chlorophyll fluorescence parameters may have contributed to better plant growth durir the PGPR treatment [38]. The growth and development of seedlings above the ground a significantly related to the root system. In this study, the root biomass was significant increased in both cucumber and pepper. Furthermore, marked differences among th control and LYGN1 treatment were found in root architecture. Beneficial microbes cou affect root development, and the altered root system architecture may improve seedlir growth by nutrient accumulation [39]. The total root surface area contributes to an increa in the total absorptive surface of the root system [40]. Additionally, total length and ro volume could also promote the absorption of more available nutrients by cucumber ar pepper seedlings. We also conducted a root diameter classification analysis by selecting range of root diameters. The results showed that LYGN1 significantly increased the ro diameter, resulting in thicker roots, especially in pepper seedlings, which may show longer lifespan and enhanced function [41]. All these indicators show the great potential LYGN1 to promote cucumber and pepper seedling growth and root development. PGP have been widely studied for improving plant growth and productivity [8,42], and th rhizosphere microbial community plays important roles in promoting plant growth an in improving tolerance to disease and abiotic stress [43]. However, there are few studi evaluating the rhizosphere microbial communities, especially under PGPR treatment, i plug seedling cultivation. *Serratia marcescens* was demonstrated to induce plant defens and improve significant growth increases in shoot length, shoot dry weight, root lengt and root dry weight [15]. Additionally, it also showed its varied beneficial traits an plant growth-promoting potential in coconut palms [19]. In this study, the α-diversity wa slightly increased after the *Serratia marcescens* LYGN1 application, but *Serratia marcescen* LYGN1 significantly changed the structure of both bacterial and fungal communities. Th obtained results from Pearson's correlation analysis showed the specific phylum whic is related to seedling growth. The correlation analysis showed that the Abditibacterio and Bdellovibrionota had positive effects on seedling growth. The relative abundance Proteobacteria were increased in both cucumber and pepper. Previous studies reported th Proteobacteria were the dominant taxonomic phyla in lettuce substrate seedlings, whic is also related to the promoting effects on lettuce growth [44]. The Proteobacteria phyl also comprise several PGPR species that promote plant growth [45], and the changes we more intense in the pepper rhizosphere, which may contribute to the growth. Furthermor Proteobacteria also showed positive correlations with different seedling growth and ro system indicators. The Firmicutes taxa have the potential to enhance plant stress toleranc growth, and nutrient uptake [46]. However, the Firmicutes were decreased in both cu cumber and pepper rhizospheres, which also showed negative correlations with differe seedling growth and root system indicators. The relative expression of Ascomycota i cucumber and pepper was decreased and showed negative correlations with the relate indicators. Previous studies showed that the initial colonization by PGPR in the rhizosphe affected the microbial community composition throughout the plant growth stages [47 which showed the potential of PGPR inoculation during seedling cultivation for promoting vegetable growth and yield.

5. Conclusions

In this study, the *Serratia marcescens* LYGN1 was used for its growth-promoting effects in plug seedling cultivation. LYGN1 improved root growth and the seedling vigor index, significantly promoting cucumber and pepper seedling growth. Our results also reveal that the microbial community composition of cucumber and pepper rhizospheres was reformed. Additionally, the correlation analysis showed that the changed microbial groups could affect seedling growth and root system development. These results show the great potential of LYGN1 in cucumber and pepper plug seedling cultivation, which could improve the growth and yield in subsequent soil cultivation. In further studies, it will also be interesting to profile the metabolites to understand the mechanism involved.

Author Contributions: Conceptualization, supervision, and writing, review, and editing, Z.C. and X.Z.; writing–original draft, X.Z. and J.P.; methodology, Y.S. and G.W.; investigation, J.P., X.H. and G.F. All authors have read and agreed to the published version of the manuscript.

Funding: This study was funded by the Zhejiang University Shandong (Linyi) Modern Agriculture Research Institute, which contributed to the local economic development project (Grant Number ZDNY-2021-FWLY02002); and the Special Project of Central Government Guiding Local Science and Technology Development (Grant Number YDZX2021070).

Data Availability Statement: The raw data for 16S rRNA gene and ITS sequences have been submitted to the NCBI Sequence Read Archive (SRA) database under the BioProject number PRJNA1051178.

Acknowledgments: The authors express gratitude to the editor and reviewers for constructive feedback and valuable comments that significantly improved the quality of this manuscript.

Conflicts of Interest: The authors declare that the research was conducted in the absence of any commercial or financial relationships that could be construed as a potential conflict of interest.

References

1. Du, X.; Si, L.; Jin, X.; Li, P.; Yun, Z.; Gao, K. Classification of Plug Seedling Quality by Improved Convolutional Neural Network with an Attention Mechanism. *Front. Plant Sci.* **2022**, *13*, 967706. [CrossRef] [PubMed]
2. Han, L.; Mo, M.; Gao, Y.; Ma, H.; Xiang, D.; Ma, G.; Mao, H. Effects of New Compounds into Substrates on Seedling Qualities for Efficient Transplanting. *Agronomy* **2022**, *12*, 983. [CrossRef]
3. Suzuki, N.; Rivero, R.M.; Shulaev, V.; Blumwald, E.; Mittler, R. Abiotic and Biotic Stress Combinations. *New Phytol.* **2014**, *203*, 32–43. [CrossRef] [PubMed]
4. Sun, X.W.; Wu, Z.H.; Feng, Y.X.; Shi, X.F.; Li, P.L.; Shang, Q.M. Research Progress on Vegetables Seedling Culture Technique during 'The Thirteenth Five-Year Plan' in China. *China Vege* **2021**, *1*, 18–26.
5. Singh, M.; Singh, J.P.; Pandey, S.K.; Mahay, D.; Shrivastva, V. Factors Affecting the Performance of Greenhouse Cucumber Cultivation-A Review. *Int. J. Curr. Microbiol. Appl. Sci.* **2017**, *6*, 2304–2323. [CrossRef]
6. Qin, Y.; Shang, Q.; Zhang, Y.; Li, P.; Chai, Y. Bacillus Amyloliquefaciens L-S60 Reforms the Rhizosphere Bacterial Community and Improves Growth Conditions in Cucumber Plug Seedling. *Front. Microbiol.* **2017**, *8*, 2620. [CrossRef]
7. Kloepper, J.W. Plant Growth-Promoting Rhizobacteria and Plant Growth Under Gnotobiotic Conditions. *Phytopathology* **1981**, *71*, 642. [CrossRef]
8. Bhattacharyya, P.N.; Jha, D.K. Plant Growth-Promoting Rhizobacteria (PGPR): Emergence in Agriculture. *World J. Microbiol. Biotechnol.* **2012**, *28*, 1327–1350. [CrossRef]
9. Shultana, R.; Zuan, A.T.K.; Naher, U.A.; Islam, A.K.M.M.; Rana, M.M.; Rashid, M.H.; Irin, I.J.; Islam, S.S.; Rim, A.A.; Hasan, A.K. The PGPR Mechanisms of Salt Stress Adaptation and Plant Growth Promotion. *Agronomy* **2022**, *12*, 2266. [CrossRef]
10. Zhou, X.; Zhang, X.; Ma, C.; Wu, F.; Jin, X.; Dini-Andreote, F.; Wei, Z. Biochar Amendment Reduces Cadmium Uptake by Stimulating Cadmium-Resistant PGPR in Tomato Rhizosphere. *Chemosphere* **2022**, *307*, 136138. [CrossRef] [PubMed]
11. Zhao, X.; Yuan, X.; Xing, Y.; Dao, J.; Zhao, D.; Li, Y.; Li, W.; Wang, Z. A Meta-Analysis on Morphological, Physiological and Biochemical Responses of Plants with PGPR Inoculation under Drought Stress. *Plant Cell Environ.* **2023**, *46*, 199–214. [CrossRef]
12. Zahir, Z.A.; Arshad, M.; Frankenberger, W.T. Plant Growth Promoting Rhizobacteria: Applications and Perspectives In Agriculture. *Adv. Agron.* **2003**, *81*, 97–168. [CrossRef]
13. Grimont, P.A.D.; Grimont, F. The Genus Serratia. *Annu. Rev. Microbiol.* **1978**, *32*, 221–248. [CrossRef]
14. Zhang, X.; Song, M.; Li, J.; Liu, X.; Gao, L.; Tian, Y. Changes in Soil Nematode and Microbial Community in Cucumber Root-Zone Soil Shaped by Intercropping with Amaranth. *Horticulturae* **2023**, *9*, 924. [CrossRef]

15. Lavania, M.; Chauhan, P.S.; Chauhan, S.V.S.; Singh, H.B.; Nautiyal, C.S. Induction of Plant Defense Enzymes and Phenoli by Treatment With Plant Growth–Promoting Rhizobacteria Serratia Marcescens NBRI1213. *Curr. Microbiol.* **2006**, *52*, 363–36 [CrossRef]
16. Niu, H.; Sun, Y.; Zhang, Z.; Zhao, D.; Wang, N.; Wang, L.; Guo, H. The Endophytic Bacterial Entomopathogen *Serratia marcescer* Promotes Plant Growth and Improves Resistance against *Nilaparvata lugens* in Rice. *Microbiol. Res.* **2022**, *256*, 126956. [CrossRe
17. Matteoli, F.P.; Passarelli-Araujo, H.; Reis, R.J.A.; da Rocha, L.O.; de Souza, E.M.; Aravind, L.; Olivares, F.L.; Venancio, T.M. Genom Sequencing and Assessment of Plant Growth-Promoting Properties of a *Serratia marcescens* Strain Isolated from Vermicompo *BMC Genom.* **2018**, *19*, 750. [CrossRef]
18. Chakraborty, U.; Chakraborty, B.N.; Chakraborty, A.P. Influence of *Serratia marcescens* TRS-1 on Growth Promotion and Inductic of Resistance in *Camellia sinensis* against *Fomes lamaoensis*. *J. Plant Interact.* **2010**, *5*, 261–272. [CrossRef]
19. George, P.; Gupta, A.; Gopal, M.; Thomas, L.; Thomas, G.V. Multifarious Beneficial Traits and Plant Growth Promoting Potenti of *Serratia marcescens* KiSII and *Enterobacter* Sp. RNF 267 Isolated from the Rhizosphere of Coconut Palms (*Cocos Nucifera* L *World J. Microbiol. Biotechnol.* **2013**, *29*, 109–117. [CrossRef]
20. Singh, R.P.; Manchanda, G.; Yang, Y.; Singh, D.; Srivastava, A.K.; Dubey, R.C.; Zhang, C. Deciphering the Factors for Nodulatic and Symbiosis of *Mesorhizobium* Associated with Cicer Arietinum in Northwest India. *Sustainability* **2019**, *11*, 7216. [CrossRef]
21. Li, H.; La, S.; Zhang, X.; Gao, L.; Tian, Y. Salt-Induced Recruitment of Specific Root-Associated Bacterial Consortium Capable Enhancing Plant Adaptability to Salt Stress. *ISME J.* **2021**, *15*, 2865–2882. [CrossRef]
22. Dong, D.M.; Wang, J.F. Relationship between greenhouse cucumber planting density and leaf area. *Chin. Veg.* **1988**, *3*, 14–17.
23. Dickson, A.; Leaf, A.L.; Hosner, J.F. Seedling Quality—Soil Fertility Relationships of White Spruce, and Red and White Pine Nurseries. *For. Chron.* **1960**, *36*, 237–241. [CrossRef]
24. Niu, B.; Paulson, J.N.; Zheng, X.; Kolter, R. Simplified and Representative Bacterial Community of Maize Roots. *Proc. Natl. Aca Sci. USA* **2017**, *114*, E2450–E2459. [CrossRef]
25. Gardes, M.; Bruns, T.D. ITS Primers with Enhanced Specificity for Basidiomycetes—Application to the Identification of Myco rhizae and Rusts. *Mol. Ecol.* **1993**, *2*, 113–118. [CrossRef]
26. Chen, S.; Zhou, Y.; Chen, Y.; Gu, J. Fastp: An Ultra-Fast All-in-One FASTQ Preprocessor. *Bioinformatics* **2018**, *34*, i884–i89 [CrossRef]
27. Magoč, T.; Salzberg, S.L. FLASH: Fast Length Adjustment of Short Reads to Improve Genome Assemblies. *Bioinformatics* **2011**, 2 2957–2963. [CrossRef]
28. Wang, Q.; Garrity, G.M.; Tiedje, J.M.; Cole, J.R. Naïve Bayesian Classifier for Rapid Assignment of rRNA Sequences into the Ne Bacterial Taxonomy. *Appl. Environ. Microbiol.* **2007**, *73*, 5261–5267. [CrossRef]
29. Quast, C.; Pruesse, E.; Yilmaz, P.; Gerken, J.; Schweer, T.; Yarza, P.; Peplies, J.; Glöckner, F.O. The SILVA Ribosomal RNA Ger Database Project: Improved Data Processing and Web-Based Tools. *Nucleic Acids Res.* **2013**, *41*, D590–D596. [CrossRef]
30. Nilsson, R.H.; Larsson, K.-H.; Taylor, A.F.S.; Bengtsson-Palme, J.; Jeppesen, T.S.; Schigel, D.; Kennedy, P.; Picard, K.; Glöckne F.O.; Tedersoo, L.; et al. The UNITE Database for Molecular Identification of Fungi: Handling Dark Taxa and Parallel Taxonom Classifications. *Nucleic Acids Res.* **2019**, *47*, D259–D264. [CrossRef]
31. Caporaso, J.G.; Bittinger, K.; Bushman, F.D.; DeSantis, T.Z.; Andersen, G.L.; Knight, R. PyNAST: A Flexible Tool for Alignin Sequences to a Template Alignment. *Bioinformatics* **2010**, *26*, 266–267. [CrossRef]
32. Caporaso, J.G.; Kuczynski, J.; Stombaugh, J.; Bittinger, K.; Bushman, F.D.; Costello, E.K.; Fierer, N.; Peña, A.G.; Goodrich, J.K Gordon, J.I.; et al. QIIME Allows Analysis of High-Throughput Community Sequencing Data. *Nat. Methods* **2010**, *7*, 335–33 [CrossRef]
33. Edgar, R.C. UPARSE: Highly Accurate OTU Sequences from Microbial Amplicon Reads. *Nat. Methods* **2013**, *10*, 996–99 [CrossRef]
34. Shi, H.; Lu, L.; Ye, J.; Shi, L. Effects of Two *Bacillus velezensis* Microbial Inoculants on the Growth and Rhizosphere Soil Environme of Prunus Davidiana. *Int. J. Mol. Sci.* **2022**, *23*, 13639. [CrossRef]
35. Chang, C.-L.; Chang, K.-P. The Growth Response of Leaf Lettuce at Different Stages to Multiple Wavelength-Band Light-Emittin Diode Lighting. *Sci. Hortic.* **2014**, *179*, 78–84. [CrossRef]
36. Chen, J.-M.; Feng, W.-M.; Yan, H.; Liu, P.; Zhou, G.-S.; Guo, S.; Yu, G.; Duan, J.-A. Explore the Interaction between Root Metabolis and Rhizosphere Microbiota during the Growth of *Angelica sinensis*. *Front. Plant Sci.* **2022**, *13*, 1005711. [CrossRef]
37. Kotoky, R.; Nath, S.; Kumar Maheshwari, D.; Pandey, P. Cadmium Resistant Plant Growth Promoting Rhizobacteria *Serrat marcescens* S2I7 Associated with the Growth Promotion of Rice Plant. *Environ. Sustain.* **2019**, *2*, 135–144. [CrossRef]
38. Pérez-Rodriguez, M.M.; Piccoli, P.; Anzuay, M.S.; Baraldi, R.; Neri, L.; Taurian, T.; Lobato Ureche, M.A.; Segura, D.M.; Cohe A.C. Native Bacteria Isolated from Roots and Rhizosphere of *Solanum lycopersicum* L. Increase Tomato Seedling Growth under Reduced Fertilization Regime. *Sci. Rep.* **2020**, *10*, 15642. [CrossRef]
39. Verbon, E.H.; Liberman, L.M. Beneficial Microbes Affect Endogenous Mechanisms Controlling Root Development. *Trends Pla Sci.* **2016**, *21*, 218–229. [CrossRef]
40. López-Bucio, J.; Cruz-Ramírez, A.; Herrera-Estrella, L. The Role of Nutrient Availability in Regulating Root Architecture. *Cu Opin. Plant Biol.* **2003**, *6*, 280–287. [CrossRef]
41. Sun, M.; Liu, X.; Shi, K.; Peng, F.; Xiao, Y. Effects of Root Zone Aeration on Soil Microbes Species in a Peach Tree Rhizosphere an Root Growth. *Microorganisms* **2022**, *10*, 1879. [CrossRef] [PubMed]

2. Nadeem, S.M.; Ahmad, M.; Zahir, Z.A.; Javaid, A.; Ashraf, M. The Role of Mycorrhizae and Plant Growth Promoting Rhizobacteria (PGPR) in Improving Crop Productivity under Stressful Environments. *Biotechnol. Adv.* **2014**, *32*, 429–448. [CrossRef] [PubMed]
3. Berendsen, R.L.; Pieterse, C.M.J.; Bakker, P.A.H.M. The Rhizosphere Microbiome and Plant Health. *Trends Plant Sci.* **2012**, *17*, 478–486. [CrossRef] [PubMed]
4. Guan, T.-K.; Wang, Q.-Y.; Li, J.-S.; Yan, H.-W.; Chen, Q.-J.; Sun, J.; Liu, C.-J.; Han, Y.-Y.; Zou, Y.-J.; Zhang, G.-Q. Biochar Immobilized Plant Growth-Promoting Rhizobacteria Enhanced the Physicochemical Properties, Agronomic Characters and Microbial Communities during Lettuce Seedling. *Front. Microbiol.* **2023**, *14*, 1218205. [CrossRef] [PubMed]
5. Bruto, M.; Prigent-Combaret, C.; Muller, D.; Moënne-Loccoz, Y. Analysis of Genes Contributing to Plant-Beneficial Functions in Plant Growth-Promoting Rhizobacteria and Related Proteobacteria. *Sci. Rep.* **2014**, *4*, 6261. [CrossRef]
6. Bokhari, A.; Essack, M.; Lafi, F.F.; Andres-Barrao, C.; Jalal, R.; Alamoudi, S.; Razali, R.; Alzubaidy, H.; Shah, K.H.; Siddique, S.; et al. Bioprospecting Desert Plant *Bacillus* Endophytic Strains for Their Potential to Enhance Plant Stress Tolerance. *Sci. Rep.* **2019**, *9*, 18154. [CrossRef]
7. Liu, Y.; Gao, J.; Bai, Z.; Wu, S.; Li, X.; Wang, N.; Du, X.; Fan, H.; Zhuang, G.; Bohu, T.; et al. Unraveling Mechanisms and Impact of Microbial Recruitment on Oilseed Rape (*Brassica napus* L.) and the Rhizosphere Mediated by Plant Growth-Promoting Rhizobacteria. *Microorganisms* **2021**, *9*, 161. [CrossRef]

MDPI AG
Grosspeteranlage 5
4052 Basel
Switzerland
Tel.: +41 61 683 77 34

Plants Editorial Office
E-mail: plants@mdpi.com
www.mdpi.com/journal/plants

www.ingramcontent.com/pod-product-compliance
Lightning Source LLC
Chambersburg PA
CBHW041343210926
51850CB00013B/2974
9783725831524